Emergent Timber Technologies

Materials
Structures
Engineering
Projects

SIMONE JESKA / KHALED SALEH PASCHA

EMERGENT TIMBER TECHNOLOGIES

MATERIALS STRUCTURES ENGINEERING PROJECTS

EDITED BY RAINER HASCHER
TECHNISCHE UNIVERSITÄT BERLIN

BIRKHÄUSER BASEL

Layout, cover design and typography
Miriam Bussmann, Berlin

Project management
Henriette Mueller-Stahl, Berlin

Translation from German into English
Philip Thrift, Hanover

Image research
Julia Pauli, Berlin

Library of Congress Cataloging-in-Publication data
A CIP catalog record for this book has been applied for at the Library of Congress.

Bibliographic information published by the German National Library
The German National Library lists this publication in the Deutsche Nationalbibliografie; detailed bibliographic data are available on the Internet at http://dnb.dnb.de.

This publication is also available as an e-book (ISBN PDF 978-3-03821-616-2; ISBN EPUB 978-3-03821-580-6) and in a German language edition (ISBN 978-3-03821-501-1).

P.O. Box 44, 4009 Basel, Switzerland
Part of Walter de Gruyter GmbH, Berlin/Boston

Printed on acid-free paper produced from chlorine-free pulp. TCF ∞

Printed in Germany

ISBN 978-3-03821-502-8

9 8 7 6 5 4 3 2 1

www.birkhauser.com

Contents

Introduction

Rainer Hascher

In the early days of building, the materials found in nature, such as wood, stone and bricks made from loam, formed the basis for all ideas about how structures should be built and what was technologically possible. Ancient models from those times still shape our ideas about buildings today. We therefore associate wood primarily with the traditional and regional forms of construction of past eras. Pictures of skilfully built timber-frame houses from the Middle Ages and delicate historic Japanese frame structures appear spontaneously in our minds. A temple complex in Japan, dating from 900 BC, is still proving its durability and functionality to this day. Highly diverse, imaginative and varied timber structures have been built by almost all cultures throughout the history of humankind.

The universal applicability for a huge range of building tasks, the technically precise, individual fashioning down to the smallest detail, robustness coupled with structural elegance and an intrinsic, natural aesthetic, which cannot be separated from the special haptic qualities, are the characteristics of this material.

Despite the extraordinarily diverse range of potential applications and its proven track record as a building material over thousands of years, wood got pushed aside during the age of industrialisation. The manual working processes of timber were too slow and hence too expensive when compared with the industrialised production methods of steel and reinforced concrete.

Recent decades, however, have seen rapid progress in the rationalisation and industrialisation of the production and processing methods for wood-based products. One of the first steps in this evolution was the transition from conventional timber construction to engineered timber structures with specific, standardised and quality-managed glued connections. This move, combined with the introduction of new jointing methods such as anchors with bolts, steel dowels, punched metal plate fasteners and steel gusset plates, made possible long spans which, in many cases, were just as economical to build as equivalent steel or reinforced concrete designs. Indeed, special loadbearing structures, e.g. shells, can even be built with a distinct economic advantage over other forms of construction with solid materials. Wood's low self-weight and its building physics benefits in the form of a low thermal conductivity plus the simple, clean working of the material, both in prefabrication and its adaptability on site, represent major advantages for many different construction tasks. Based on targeted research, it has been possible to standardise the demands placed on the quality of the material to a large extent and to optimise the building technology properties of wood to suit specific requirements.

Even now, new wood-based materials still appear on the market at regular intervals. These products have attributes that were originally reserved for other materials and in some cases achieve almost isotropic behaviour. Wood, that natural product exhibiting non-uniform growth and non-uniform properties, is increasingly becoming a totally industrially manufactured material with characteristics that can be predetermined exactly.

We are currently witnessing an ongoing renaissance in timber construction, which can be principally attributed to two unrelated causes:

On the one hand, wood has been rediscovered as one of the most important renewable raw materials for sustainable building. Wood is readily available locally in many regions and can be used for construction, as a general material and as fuel. Trees absorb water and carbon dioxide (CO_2) and use solar energy (photosynthesis) to convert those substances into pure oxygen and water, which are released into the environment. As products, trees supply timber, which consists of 50% carbon; one tonne of dried construction timber contains 510kg of carbon, which corresponds to 1.8t CO_2. This CO_2 remains bound in all wood products. Only when it rots or burns is exactly this amount of CO_2 released into the atmosphere again. Wood is therefore a "zero-carbon" material and can be fully returned to the ecological cycle without producing any non-degradable wastes or residues. Compared with other building materials, the additional energy required for production and processing is extremely low and so wood is an almost ideal example of the "cradle-to-cradle principle".

On the other hand, constant further developments in the use of three-dimensional CAD models coupled with fully prepared production data for digitally controlled robot production has been causing a revolution in engineered timber structures over the past few years. Especially interesting for architects and engineers is the fact that CNC (computerised numerical control) machines achieve not only maximum precision components with this comparatively easily machined material and the use of modern control technology, but are also ideal for the automated production of complex geometries and structures which are far superior to those produced by mechanically controlled machines in terms of accuracy and speed of production. An additional advantage is that constant human supervision of the production process is often unnecessary, because the controls provide sufficient options for integrating even quality-control methods fully automatically into the production process. This opens up totally new design options for architects and engineers: intricate, bespoke components can now be fabricated economically without the need for large batches; new custom designs – from large-scale forms down to industrially manufactured decoration – are suddenly possible. The economic approach to the consistent, rigid, orthogonal design principle for a building is losing its importance, and more elaborate structures based on more liberal design principles are becoming possible even in the face of tough economic constraints. Traditional wood joints, once the province of carpenters and joiners and recently ousted by steel connectors, are increasingly becoming relevant once more, even considering the economics, and long since forgotten methods such as beech dowels, scarfs, tenons, etc., are modern again.

Hence, there are many good reasons to assume that timber, as one of the oldest building materials in the world, will become more important in the future as a construction material in sustainable forms of construction. This book is intended to encourage ideas for thinking about wood as a building material in a new way, developing existing methods of timber construction creatively and discovering innovative design concepts on the basis of new methods and technologies.

Following a basic introduction to timber as an efficient building material, the earlier forms of timber construction and technological advances, the majority of the chapters in this book are dedicated to these new approaches. Besides the ongoing improvements to engineered connections and the introduction of digital production methods, it is the innovative development of this material that we find on the following pages. In particular, the sections on composite forms of construction with steel, textiles, concrete and glass plus the experimental structures indicate the trends that point the way forward for construction with timber. The recently completed projects described provide typical examples and details of the innovative approaches and present the reader with a direct link to practice.

Historic design typologies

Khaled Saleh Pascha

Timber beams and joists have always been used as building materials for suspended floors. Tree trunks can be obtained from forests with little effort. Working them to form loadbearing members does not require any demanding technology.

The simple roof structure consisting of just a series of parallel joists was the earliest and simplest form of roof construction. Until well into the 20th century, timber joist floors remained the standard form of construction for suspended floors. Only with the arrival of reinforced concrete in buildings in the 1920s did steel beams and concrete slabs start to be used in addition to the otherwise customary timber joist floors.

The length, width and depth of the timber beam, or the dimensions of the tree trunk, inevitably determine its usage and the maximum span of the structure. Where the beams cut from a tree trunk are not adequate, then composite (or compound) sections represent one option for creating larger and longer beams. Both deeper beams (for heavier loads) and

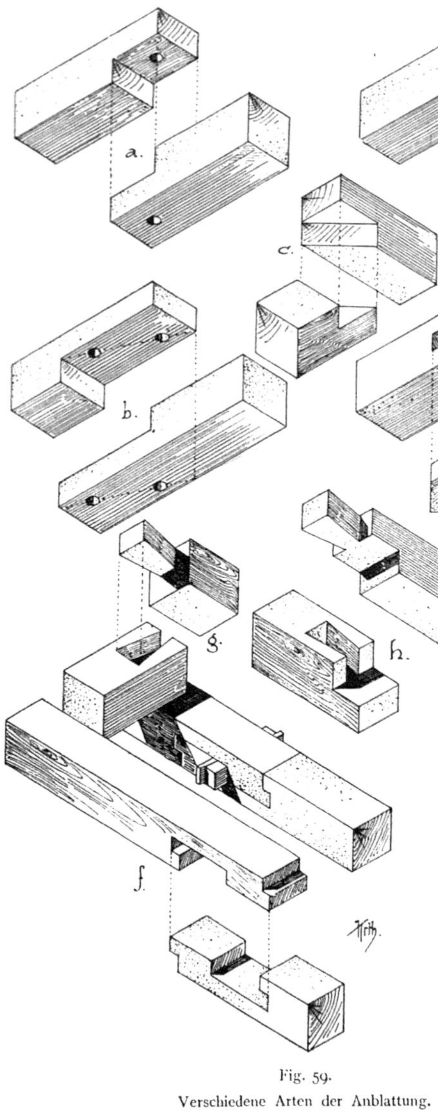

Various types of lapped joint. Such a wood joint represents one option for transferring tension between two timber components; source: Krauth, Theodor; Meyer, Franz S.; *Das Zimmermannsbuch*; Leipzig 1895; reprint Hanover 1981, p. 83.

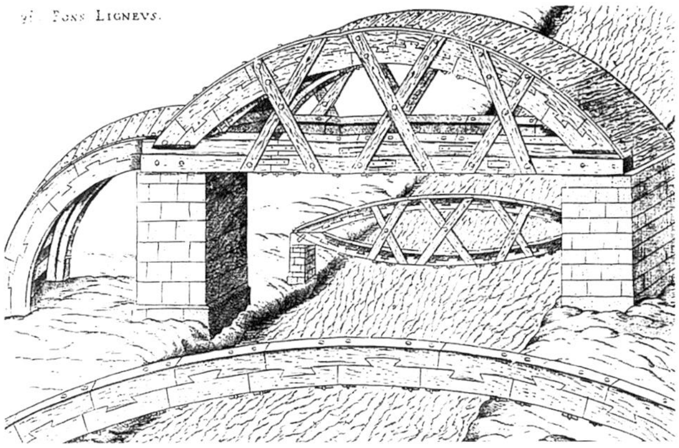

Roman timber bridge; source: Graubner, Wolfram; *Holzverbindungen*, p. 94.

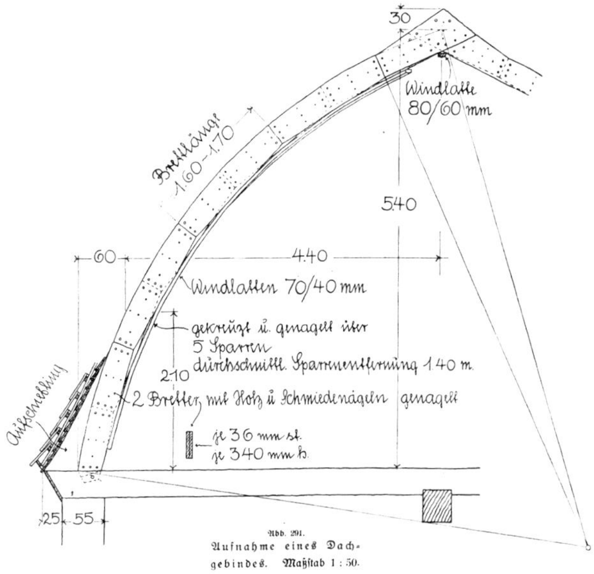

Traditional construction of a curved plank roof; source: Steinmetz, Georg; *Grundlagen für das Bauen in Stadt und Land*, Vol. 2: "Besondere Beispiele"; 1917, p. 217.

longer beams (for greater spans) are possible in this way.

In their simplest form, composite timber beams are simply joined together with ropes made from bast fibres. Such a method of jointing cannot achieve a truly shear-resistant connection between the individual elements, i.e., transferring shear forces, bending moments and torsion is possible to a very limited extent only. As a result of this non-shear-resistant connection, the compound cross-section – in terms of load-carrying capacity and deflection – behaves as though the individual beams were placed next to each other. Larger and longer beams are possible, but in structural terms they never achieve the quality of one solid cross-section.

Labour-intensive wood joints such as scarfs, dovetails and laps in various, sometimes elaborate, forms can be used to achieve a secure, structural connection between the individual beams so that their loadbearing behaviour is comparable with that of a solid beam.

The illustration of a Roman timber bridge shows the principle of shear-resistant interlocking to create structural connections between the separate pieces of a timber arch. Composite beams can be seen in both the arch and the deck. Those in the arch employ dovetail joints and dowels to achieve a shear-resistant connection, those in the deck use splice (or fish) plates, which are less effective structurally.

It was around 1561 that the architect to the French court, Philibert de L'Orme, developed a composite member made up of curved planks, which, linked together with cross-members running longitudinally with respect to the building, meant that long spans were now possible. This new type of composite arch member was first mentioned in his book *Nouvelles inventions pour bien bastir et à petits fraiz*, in which he describes a laminated type (usually 2–3 plies) of construction with vertical planks which could be used to build long-span arches.

In Germany the composite member with curved planks became particularly popular after the appearance of David Gilly's book *Handbuch der Landbaukunst* in 1798. Compared with the conventional couple roof, this form of construction resulted in lower, even zero, shear forces at the supports due to its barrel-vault form. The advantage over the purlin roof is the saving in materials and the fact that there are no intervening columns.

The disadvantages of the composite member with curved planks are primarily its time-consuming fabrication and the high wastage when cutting the curved laminations to size. Owing to the high number of joints compared with a solid beam, the bending stiffness at the joints between individual laminations for a two-ply form is reduced by half. Despite all these drawbacks concerning the structural performance of the construction, long spans were possible. For example, a span of 41m was achieved as early as 1783 for the roof to a grain store in Paris.[1]

One particularly interesting variation on the composite timber beam is the interlocked beam developed by the Grubenmann brothers, the Swiss bridge-builders and master carpenters. They used this for Limmat Bridge at Wettingen Monastery in Switzerland (1765–1766). Spanning 61m, this bridge was at the time an engineering masterpiece.

Another variation on the composite beam is the Emy form of construction, which is named after its inventor, the French engineer Armand-Rose Emy (1771–1851), who developed this form around 1830. This is a laminated beam without any adhesive which has horizontal planks clamped together by means of wrought bolts and iron straps let into the timber. The main difference between this and the composite form of Philibert de L'Orme is that Emy's invention has stacked laminations fixed together, whereas De L'Orme uses upright laminations. Very long spans are possible with the horizontal laminations of Emy's form of construction.

A fundamental risk is that high flexural stresses can lead to individual laminations becoming displaced or dislodged from the line of action of the forces because the clamped and bolted connections are at isolated points only.

When using composite members, the forces acting on the structure are transferred across the (offset) joints between individual segments by the other segments continuous at that point. In the 1920s, this principle prompted Fritz Zollinger, city building surveyor at Merseburg near Leipzig, to join together the individual board or plank laminations in a diamond pattern to form a lattice shell.

1 Another example of an early composite member with curved planks was Paretz Palace in Brandenburg (1797, architect: David Gilly).

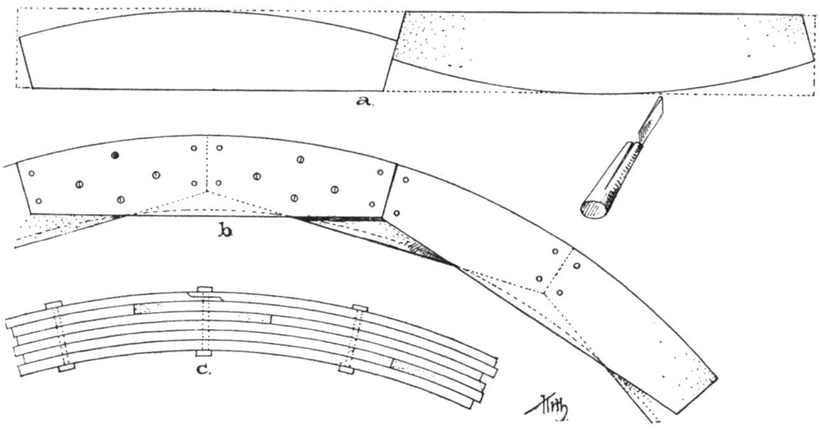

Fig. 154.
Konstruktion der Bohlenbogen.

Fabrication of an arch with curved planks around 1910; source: Krauth, Theodor; Meyer, Franz S.; *Das Zimmermannsbuch*; Leipzig 1985; reprint Hanover 1981, p. 175.

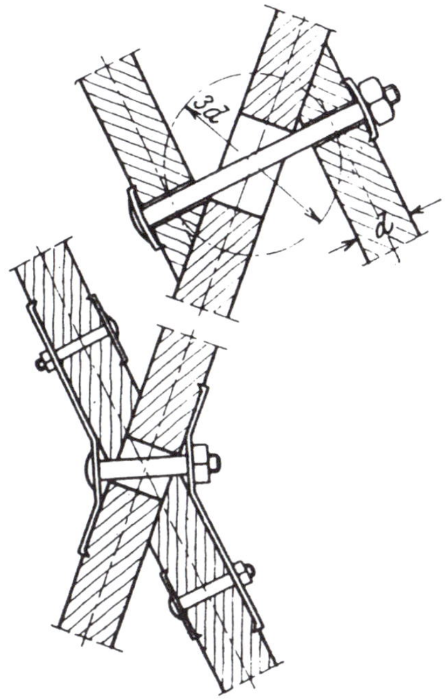

The principle of the Zollinger form of construction: individual boards or planks are joined together in a diamond lattice to create a shell structure.

A contemporary example of the Zollinger form of construction. The development of the Bertsche anchor has resulted in a non-slip, rigid connection between the individual laminations. HanseMesse Rostock, 2002; architects: Gerkan, Marg & Partner; engineers: schlaich bergermann & partner.

The use of identical timber laminations in single curvature, pairs of which are bolted together at an angle to a continuous lamination at the same level, produces a diamond-shaped lattice that can be used for diverse roof forms including round, segmental and pointed arches.

The disadvantage of the Zollinger form of construction is the relatively high deformation of the structure due to the non-rigid connections between the laminations where they intersect. As the connections between individual laminations are not on the line of action of the forces but marginally eccentric due to the form of construction itself, after a time, following the inevitable shrinkage of the timber perpendicular to the grain, the timber laminations experience slip at the nodes, which over the long term can lead to large deformations and sagging of the structure.

A number of contemporary shed structures (exhibition halls in Rimini, 2001, Rostock, 2002, and Friedrichshafen, 2002) demonstrate that, provided the problem of the minimal rigidity at the connections can be solved, the Zollinger form of construction is, even today, a cost-effective, efficient and aesthetically appealing option for barrel-vault shells.

In the 1830s and 1840s, the architect, engineer and town planner Georg Ludwig Friedrich Laves developed a truss with a fish-belly (lenticular) form, the "Laves beam". This form matches the bending moment diagram for this type of member. The primary element in this form of construction is a timber beam split down the middle lengthwise and splayed apart, which increases the structural depth at mid-span and makes it considerably more efficient than the original timber beam. In order to resist the high tensile stresses perpendicular to the grain, the ends of the beam are strengthened with bolts, which prevent the beam from splitting apart completely. One variation consists of two beams connected at their ends and splayed apart at mid-span. By dividing the beam into top and bottom chords with intervening struts, which determine the spacing of the chords and also transfer the forces between them, this beam anticipates many elements of modern truss construction.

Trussed timber beams employ diagonals in tension (e.g., wire ropes) below the plane of the beam or diagonals in compression (e.g., timber members) above. Such framing above and/or below a timber beam reduces its deflection and therefore upgrades it considerably in structural terms. These forms of construction should also be understood as the forerunners of the truss in which diagonals in tension (ties) and/or compression (struts) are included depending on the loads on the structure.

If, as in the case of roofs, there are no constraints on the depth of a structural element (in contrast to suspended floors), then framing above or below the level of the beam itself, or combinations of the two, as well as other truss forms represent a reasonable strategy for achieving longer spans. The great advantage of these forms of construction is that the individual members, whether horizontal, vertical or diagonal, can be made from single elements, which in most cases, even with very long spans, are easily fabricated from individual, solid timber sections with small dimensions. The good structural efficiency of this form of construction is achieved by the resolution into compression and tension members, the top and bottom chords, which are joined together by diagonals and zero-force members. As, in the ideal case, all members are subjected to axial forces only, it is possible to create efficient timber structures in which the material is used ideally.

Whereas, as in the case of the historical trussed beams for traditional roof structures, spans of 20m were possible as early as the 16th century[2], the development of modern timber truss structures achieved a quantum leap forward in the 19th century in terms of their spans and the efficiency of their designs. One example is the shed for the 4th German Choirs Festival in Vienna in 1890, the roof beams of which spanned 56m. Knowledge gained from the modern steel trusses of the 19th century, especially the structural clarity of the individual elements and the connections developed from structural steelwork, was transferred back to timber truss design and is crucial to modern timber engineering.[3]

2 See A. Gattnar, F. Trysna, *Hölzerne Dach- und Hallenbauten*; 7th ed., 1961, p. 8.

3 See A. Gattnar, F. Trysna, *Hölzerne Dach- und Hallenbauten*; 7th ed., 1961, p. 10.

For towers too, modern timber truss forms with new steel connectors and their numerical analysis render possible dimensions that were inconceivable hitherto.

Representative of a whole series of towers and masts built in timber in the early 20th century is Gliwice radio tower, a 118m high timber truss made of larch erected in 1935. The influence of the steel truss structure of the Eiffel Tower in Paris is clearly recognisable. The individual columns, beams and diagonals of Gliwice radio tower are themselves resolved into truss forms. Furthermore, due to the different structural depths of the beams, similar to the case of the Eiffel Tower, the intersection points of the elements are separated, which enables collar-type connections, with continuous intermediate members, which in themselves represent structural and constructional advantages and are crucial for the extreme delicacy of the construction.

The Bertsche anchor is a metal connector between the individual laminations of the glulam construction. The offset connection point between the laminations, at approx. 60° to the plane of the beam, is readily apparent.

Gliwice radio tower, Poland.

New technologies and methods

Simone Jeska

Effective connections, the material properties of wood and the choice of loadbearing structure are all fundamental to efficient timber structures. If, in addition, we consider timber structures to be competing with those made from steel or reinforced concrete, then wood's anisotropy, its brittle failure behaviour and the need to use steel to join the elements to form a loadbearing system are the main negative factors when it comes to timber structures. Accordingly, many current research projects are concerned with developing new connections, establishing new forms of loadbearing construction and new materials, strengthening connections and upgrading materials.

Connections

Mechanical connections with steel

Connections in timber construction can be in the form of traditional wood joints or make use of mechanical steel connectors or be glued. Self-drilling screws, steel fasteners fixed with adhesive, jumbo corrugated fasteners or the highly efficient Sherpa system connectors and Bertsche BVD system represent new types of steel connections for timber construction which have established themselves over recent years. Self-drilling screws are increasingly taking the place of the old connections with sheet metal hangers. Screws are also used to form rigid connections at frame corners and to attach diagonal members to the top and bottom chords of trusses[1]. With hardened shafts and special tips, these screws can be driven directly into wood or wood-based products using special tools with high torques and therefore represent a relatively straightforward way of producing structurally effective connections on site. The high tension capacity (pull-out resistance) of the screws is exploited to join the components together by driving them into the timber at an angle (permissible angles are currently 45°–90°). As the angle increases, the stiffness and load-carrying capacity of the connection also increase in proportion to the driving angle. The stiffness of a connection with screws at an angle of 45° can reach 12 times that of a connection with screws at 90°; the load-carrying capacity is also much higher. With several parallel, diagonal screws the stiffness and load-carrying capacity are increased yet further because the screws generate pressure between the components, which in turn generates a frictional force that can be taken into account in the design.

1 In 2009 the 41st training course organised by SAH, the Swiss working group for timber research, included a presentation of the research carried out by Hans Joachim Blaß. His proposal was to use screws or threaded rods for the connections between truss chords of glulam and diagonals of cross-laminated timber.

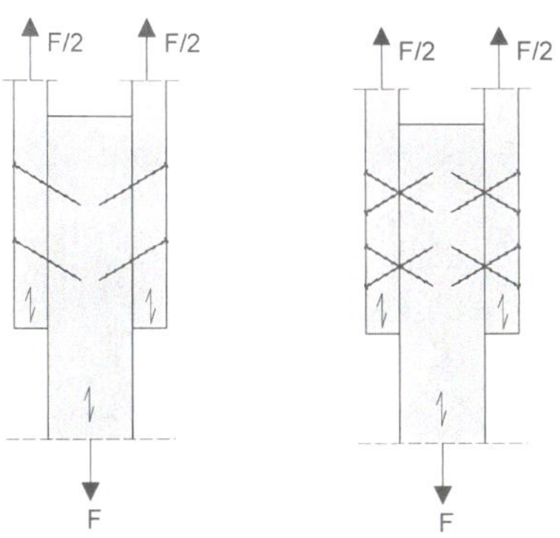

Diagonal self-drilling screws, parallel or crossing, can transfer forces acting parallel to the joint.

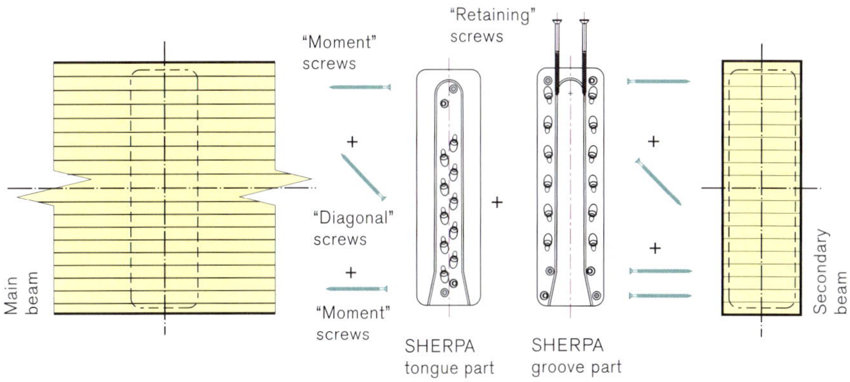

The Sherpa system connector with milled dovetail groove and matching tongue can accommodate tension, compression, shear and torsion.

Arranging the screws to cross each other resolves the loads into diagonal tensile and compressive forces; for example, at a beam-column connection, one screw is loaded in tension, the other in compression[2]. Parallel crossing screws can even be used to join components that are not in direct contact with each other, but have to be connected, even though they are separated by a cavity or soft intermediate layer (insulation).

Other new mechanical connections in engineered timber construction use the two-part connector systems well known in furniture production; these are now also available as heavy-duty connectors for joining primary and secondary beams. Such connections consist of shaped aluminium or steel parts with milled tapering grooves or elongated holes in one part and matching pins or tongues on the other part. Both parts are fixed with special full-thread screws to the timber components so that the beams can be simply slotted together[3]. Driving the screws at an angle to the grain results in highly efficient connections with a characteristic load-carrying capacity of up to 280kN.

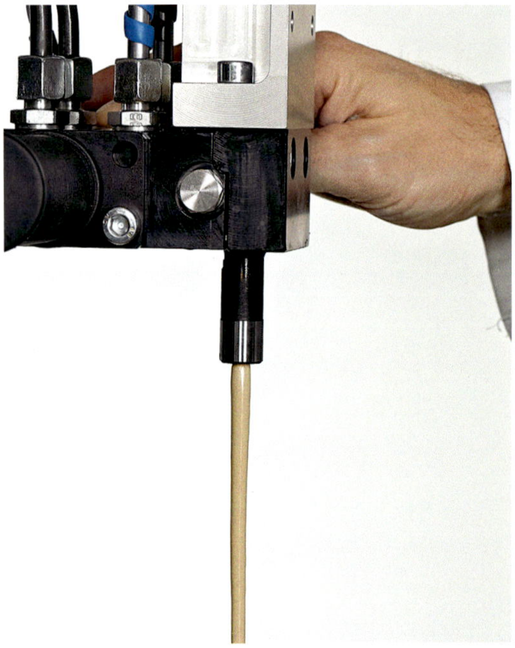

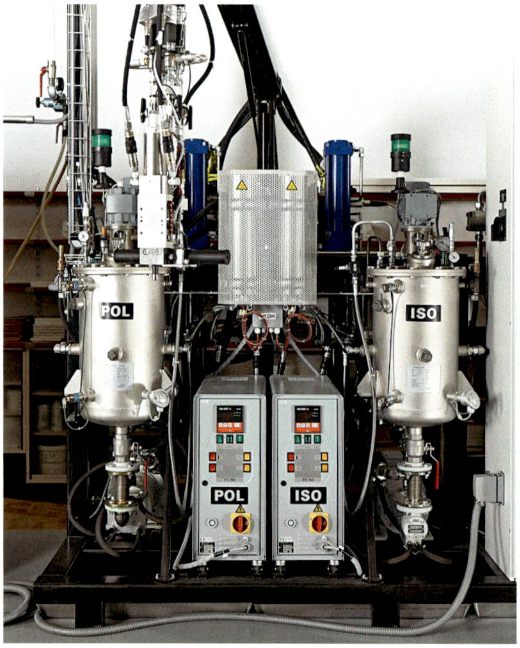

The "Lignofast" two-part adhesive can be applied by spraying or pouring. The properties of the adhesive (hardness, elasticity, colours, etc.) can be specified and set individually. Special mixing and dosing plant is required to apply this fast-reaction adhesive. The high-pressure counter-current method is well known from the automotive industry.

Glued connections

In principle, glue joints represent the most efficient way of forming a structural (i.e., force-transferring) connection between two elements. The introduction of glue joints in the production of glued trusses more than 100 years ago was a quantum leap for timber construction. Adhesives underwent further development and improvements, which in turn enabled the production of the efficient glued laminated timber beams so prevalent in engineered timber designs today. Melamine resin, polyurethane, resorcinol resin and epoxy resin are the substances normally used as effective adhesives for timber structures. New developments are often modifications of these known systems, carried out to deal with specific requirements.

The use of "high-performance adhesives" has brought about a further change in timber construction. One-part polyurethane adhesives – which are applied in a thin layer, cure at room temperature, are waterproof, have open times[4] of 20 minutes, short press times of just 15 minutes and very short setting times – not only cut production times, but also enable glue joints to be produced on site. For example, the scarf joints in the golf clubhouse roof in Yeoju, South Korea, were glued on site with PUR prepolymers. Another new development is a two-part polyurethane adhesive that sets in a few seconds and thus reduces production times to the absolute minimum. This ultrafast-setting adhesive is processed with dynamic low-pressure mixers or using the high-pressure counter-current method so that, in future, wood-based materials can be manufactured at a rate equal to that of industrial production[5]. Using high-performance adhesives improves productivity and also enables timber beams to be produced on demand. This not only saves storage costs, but enables contemporary architecture, with its double-curvature geometries and loadbearing structures made from individually shaped components, to be realised within an economically viable cost framework.

The use of elastic adhesives in the production of glued laminated timber extends the range of applications and increases efficiency. Bern University of Ap-

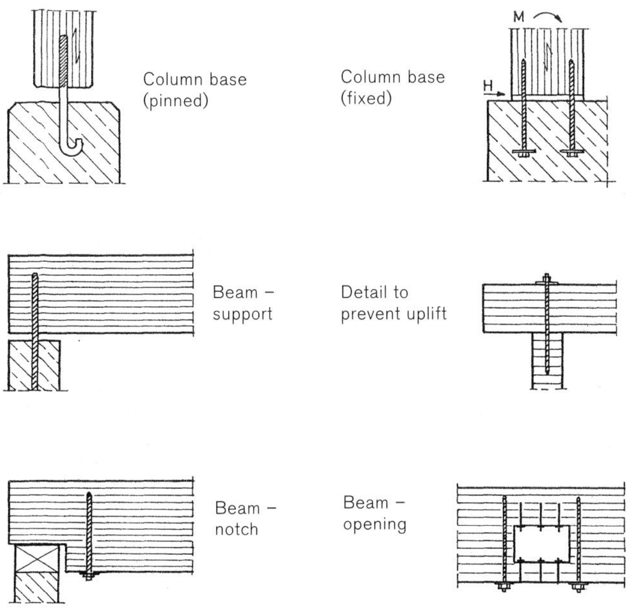

Examples of the use of glued threaded rods in timber construction of the early 1980s.

2 *Selbstbohrende Schrauben und ihre Anwendungsmöglichkeiten*, H.J. Blaß, I. Bejtka, *Holzbaukalender*, Karlsruhe 2004.

3 It was in 2009 that Harrer GmbH developed the Sherpa system connectors together with holz.bau forschungs GmbH from Graz University of Technology. Similar connectors are produced by the Simpson Strong-Tie and Knapp companies.

4 The production of glue joints is divided into three phases: 1) open time is the interval between applying the adhesive and joining the laminations together; 2) closed time is the interval between joining the laminations together and applying the clamping pressure; 3) press time. Up until now, the long press times needed for the production of glulam beams have often led to hold-ups during production. Polyurethane adhesives with a 5min open time and 2min press time have recently become available.

5 The new two-part adhesive was patented in 2008 by the Nolax company under the name of "Lignofast". This fast-reaction polyurethane adhesive was used for the first time in April 2012 to produce plywood boards. However, as conventional production machinery cannot process this fast-reaction adhesive, it is not yet possible to manufacture glulam members with the new adhesive.

plied Sciences, in cooperation with the adhesives industry, has developed glued laminated timber beams with an elastic glue joint in the tension zone of the beam; the adhesive was designed to match the requirements exactly. With a flexural capacity about 20% higher and stiffness about 80% lower when compared with a glued laminated timber beam in which the laminations are rigidly bonded, these beams are suitable for structures in seismic regions and for long-span structures where deformations play a subsidiary role[6].

Glued connections with dowels and plates

At the same time as new adhesive systems for timber construction were being developed, glues were also being increasingly used as new loadbearing connections in engineered timber structures. In principle, glued connections can be divided into joints with dowels and plates made from steel, timber or polymers glued into or onto the members and profiled joints joined together directly, e.g., finger or scarf joints.

The advantages of timber/steel glued connections over pure mechanical connections have long since been exploited in the form of glued threaded steel rods to create simple junctions at supports or to create beam splice joints[7].

This rigid and at the same time ductile connection is easy to assemble and cost-effective. Furthermore, the cross-section at the junction is not weakened so severely and this has a positive effect on the sizing of the timber members. In addition, by shifting the position of the joint, say, into the middle of the timber components, the connection becomes invisible. Over recent years, many new timber/steel glued connections have become established. Glued threaded rods represent a new type of dowelled connection in engineered timber construction and are used to create high-performance nodes in long-span gridshells as well as rigid and hinged frame corners or column bases for large single-storey sheds. One example of the use of this type of connection is the relatively small timber dome (d = 14.00m) to the

6 Maurice Brunner and Marc Donzé from Bern University of Applied Sciences in Biel, Switzerland, developed the glulam beam with elastic joints together with industry.

7 Since the mid-1970s, research projects have been carried out to investigate the form of connections using glued threaded rods; Hilmer Riberholt, Technical University of Denmark, 1973 / Karl Möhler, Klaus Hemmer, Karlsruhe Institute of Technology, 1981. Despite the absence of standardised design methods, this connection method has been regarded as state of the art since about the mid-1990s.

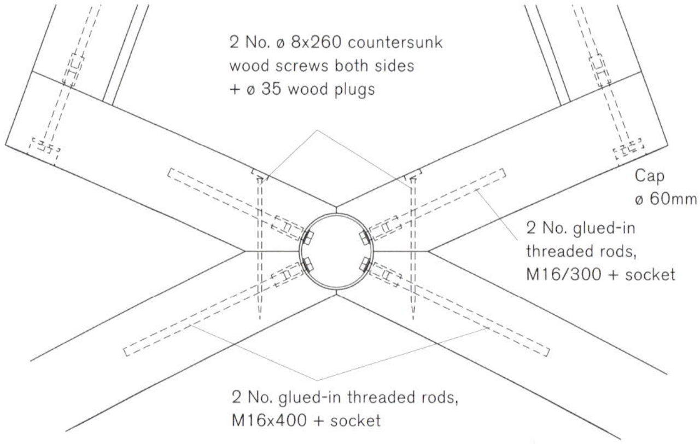

The dome in Aichi, Japan, is made up of curved glued laminated timber ribs with two different radii. The rigid nodes make use of glued threaded rods, Burgbacher company, 2005.

With a span of 120m, the timber dome over the salt store in Rheinfelden, Switzerland, shows the potential of modern engineered timber structures. Glued-in steel components connect the short glulam members of the shell together at the nodes, Häring & Co. AG, 2012.

Turkish Pavilion at Expo 2005 in Japan[8]. Four curved ribs with threaded rods glued in their end faces meet at every node in the dome. Each threaded rod has a threaded socket at the end so that a steel tube (d = 152.4mm) can be connected with M16 bolts to form the node. The metal connections in the dome are concealed afterwards by round timber caps.

Despite the lack of national technical approvals for connection methods with adhesives, the high degree of efficiency of such glued connections has seen them used more and more since the turn of this century. The efficiency of this new type of connection is demonstrated impressively by the gridshell domes to the salt store in Riburg, Switzerland (120m span), and the sports centre in Scunthorpe, UK. The gridshell in Scunthorpe, which actually consists of five intersecting domes with heights of up to 20m and diameters of up to 65m, is made up of straight glued laminated timber members (GL 32c[9]) 600m deep and 160–200mm wide arranged to form triangles. Six timber members with glued threaded rods (M16/M20) parallel with the grain meet at each welded node element, where they are simply bolted together[10]. The production of the timber/steel adhesive joint under controlled factory conditions involved forcing the epoxy resin into the 350–550mm deep drilled holes through drilled filler holes using a compressed-air gun. Drilled vent holes ensured that the holes for the threaded rods were filled completely.

In addition to connections with glued threaded rods, the potential of glued steel plates and tubes as

Glued-in tubes for tests at the Building Materials Testing Institute in Wiesbaden, Prof. Leander Bathon.

There are four steel tubes glued into the end face of each glulam beam and connected to the steel node with bolts. The oversized "football" of timber members was built on the premises of the Haas company in 2006.

structural connections in timber construction has also been investigated over recent years. The first pilot projects and experiments have revealed the efficiency of these connections. Tensile tests on specimens with four glued steel tubes resulted in failure of the connection at 450kN. At that load, the stresses in the specimen reached 1.5 times the characteristic strength of customary glued laminated timber beams (GL 24). The steel tubes, known as HSK[11] pipe connectors, are 50mm in diameter and 125mm long. To improve the bond, the outside of the tube has a double thread. There is an integral M16 thread at the end for a bolt. This connection was tried out for the first time on a temporary timber structure consisting of timber members arranged as hexagons and pentagons to form a sphere with a diameter of 26m.

In the meantime, various systems with glued steel strips or perforated plates have become established for the heavily loaded, rigid corners to single-storey sheds. GSA connectors[12] consist of steel strips glued into the tension and compression zones of the columns and beams at frame corners. A threaded bar connects the glued-in strips and resists the resulting transverse tension. Formed as a tongue and groove joint, on site, the components only need to be slotted together and secured with bolts.

The HSK frame corner works according to a similar principle. In this case custom-made steel parts, consisting of a steel flange with three parallel, welded perforated plates and welded eye-bars, are glued to the inside and outside of the beam and column such that on site the beam and column only have to be slotted to-

The GSA hinged connection consists of two half-shells that are pulled together via a conical ring and a bolt.

8 Since DIN 1052 was revised in 2004, it has been possible to design connections with glued threaded rods.

9 "GL" stands for glued laminated timber, the number denotes the permissible characteristic flexural strength in N/mm^2 and "h" stands for homogeneous, which means that all the laminations of the glulam beam comply with the requirements of the given strength class.

10 The design of the connections was carried out on the basis of experiments performed at the University of Bath and in the laboratories of timber fabricator Mayr-Melnhof Kaufmann Reuthe GmbH.

11 HSK is the German abbreviation for timber/steel glue joint. It was tested at the Building Materials Testing Institute in Wiesbaden and the timber construction laboratory at the University of Applied Sciences Wiesbaden Rüsselsheim.

12 GSA is the German abbreviation for thread-bar anchor; this highly efficient connection technology was developed by Prof. Ornst Gehri together with the Swiss company Neue Holzbau AG.

The local strengthening in the form of ash laminations glued in place enables the frame corner connection with GSA connectors to achieve a degree of efficiency of 1. With a flexural strength of $40N/mm^2$, the load-carrying capacity of grade GL 36 glulam beams can be fully utilised. EIZ (maintenance and intervention centre), Frutigen, Switzerland, Müller & Truniger Architekten, 2007.

gether and joined by a bolt to create a rigid joint. The glued perforated plates ensure that the forces from the timber cross-section are transferred uniformly to the steel flange. With simple assembly on site and the ability to transport large timber frames in separate pieces, this new type of connection is perfect for applications in single-storey sheds on remote sites with difficult access.

Attempts have been made to establish timber dowels as glued connections in engineered timber structures as an alternative to glued-in threaded rods. Structural forms such as the Zollinger system, made up of short timber members, are especially suited to this method of connection. The engineers for the thermal baths in Bad Orb and Bad Sulza adapted the Zollinger form of construction so that they could use free-form lattice shells[13] to span over spacious interiors. The lattice shell is made up of 160 × 240mm glued laminated timber members about 3.60m long arranged on a 1.80m square grid. Transferring shear forces at the nodes between the curved timber members is achieved solely by beech dowels (30mm diameter, about 14cm long) glued in on one side. A small steel plate nailed in place secures the node during erection. This simple, minimalistic connection was possible because the grid

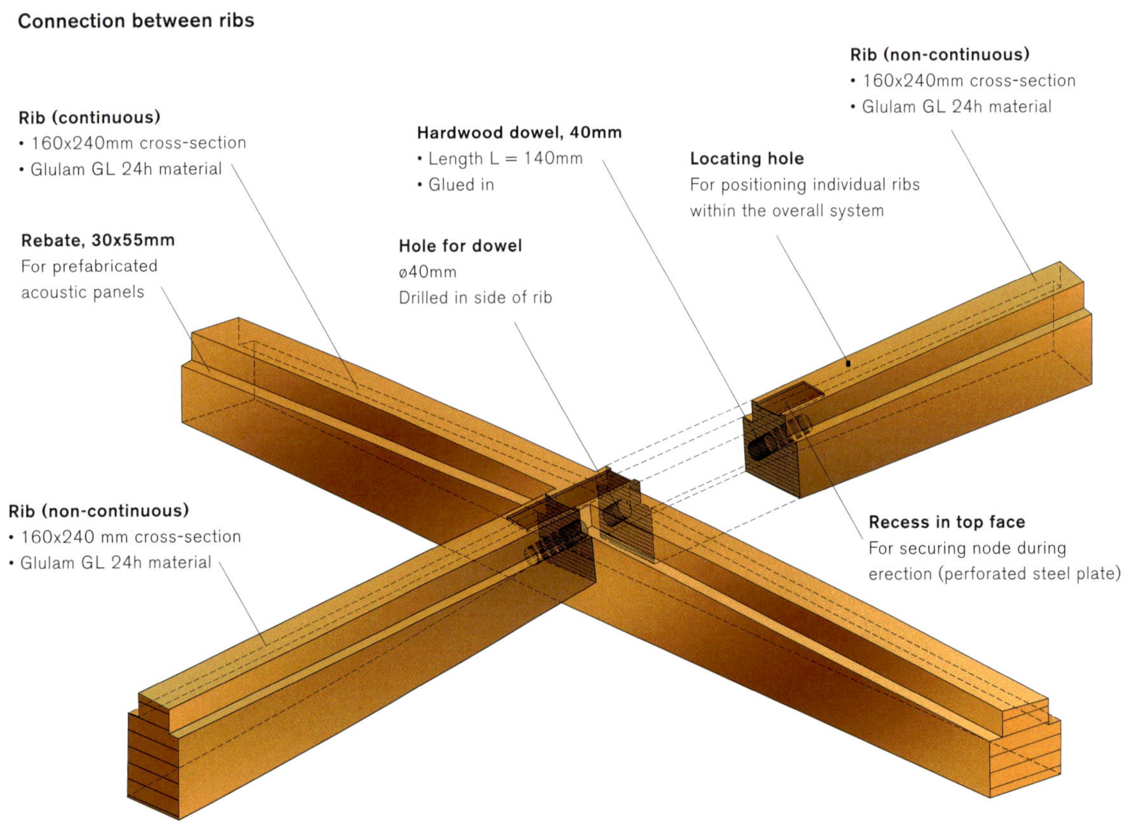

Glued-in timber dowels form the connection between the glued laminated timber members at the nodes. Toskana thermal baths, Bad Orb, Germany; architects: Ollertz Architekten; structural engineers: Trabert + Partner, 2010.

of ribs, as the third layer, forms a shell structure together with the two layers of shear-resistant timber sheathing on top.

For more than 10 years, researchers in Japan and New Zealand have been investigating butt-jointed glued laminated timber beams with glued hardwood or thermally compressed timber dowels[14]. The joint is in the form of several staggered rows of timber dowels in the tension and compression zones of the beam. The timber dowels are glued into the members parallel with the grain of the wood using a one-part polyurethane adhesive. Several series of tests were carried out on this type of connection using 12mm diameter sugar maple dowels in different arrangements in glued laminated timber beams made from Japanese cedar[15]. The results of the tests show that the load-carrying capacity of the connection is comparable with that of customary steel dowel connections, although the brittle failure behaviour of the timber dowels is a disadvantage.

For connections loaded exclusively in tension, glued-in fibre-reinforced polymer rods could offer an alternative to steel rods. The strength of CFRP (carbon fibre-reinforced polymer) rods in the longitudinal direction is higher than that of steel rods. In addition, they do not corrode, are easy to cut to size and install and are absolutely excellent for combining with timber. Studies carried out at the University of Kassel have revealed the suitability of these connections for practical applications. The background to the research was attempts to upgrade old timber structures by subsequently converting the traditional wood joints into tension-resistant connections[16]. Experiments and numerical analyses have verified the plastic rods (10–12mm diameter, 50–200mm long and glued in place with a modified two-part epoxy resin adhesive) as suitable connectors. However, long-term studies, which would enable the rods to be used in practice, are still lacking[17].

Glued wood joints

Every additional connector introduced into a timber structure weakens the timber cross-section and damages the fibres of the wood, which has a negative effect on the dimensioning of the components. Gluing timber components together directly enables the form and cross-section of the timber to be preserved and, in addition, allows the two-dimensional adhesive joints to transfer the forces uniformly over a large area. A lack of design methods and data on the adhesives as well as

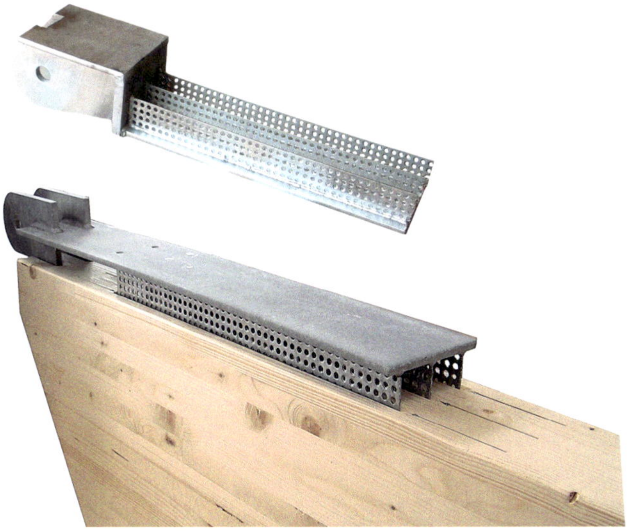

HSK frame corner: The perforated plates of the steel parts are factory-glued into the tension and compression zones of the frame corner. On the outside they function as a tie, on the inside as a strut. The frame corner was tested for the first time on the three-pin glulam frame (12m span) to the "Bergehalle" near St. Moritz, Switzerland.

13 The form of the double-curvature roof surface follows a catenary curve, the natural curve of a suspended rope. Designed as the struts to a shell, exclusively normal forces at the nodes is the result.

14 Kohei Komatsu, Jorgen L. Jensen, Akio Koizumi and Takanobu Sasaki, in various teams, have been investigating high-performance connections with glued hardwood dowels since 1997.

15 *Moment-resisting joints with hardwood dowels glued-in parallel to grain*, J. Jensen, T. Sasaki, A. Koizumi, Institute of Wood Technology, Akita Pref. University, WTCE, 2004.

16 The use of GFRP (glass fibre-reinforced polymer) rods in the refurbishment of old timber structures and for replacing damaged timber components, the so-called BETA method, has long since been standard practice.

17 See *Untersuchungen zum Verbund zwischen eingeklebten stiftförmigen faserverstärkten Kunststoffen und Holz*, Carsten Pörtner, dissertation in the publication series *Bauwerkserhaltung und Holzbau*, No. 2, Kassel, 2006. Research into connections with glued fibre-reinforced polymer rods has been ongoing since 1991 (Müller, von Roth, Hollinsky). However, owing to further developments affecting both polymers and adhesives, the early research results are only partly relevant.

adhesive systems suitable for use on site have so far prevented the introduction of structurally effective, two-dimensional adhesive joints for junctions and nodes. One difficulty when determining the load-carrying capacity of adhesive joints is the non-linear stress distribution in the adhesive joint itself. When gluing overlapping, elastic components, stress peaks occur in the joint at the ends of the overlap, which leads to failure of the connection.

The latest studies have proved, however, that the load-carrying capacity of glued connections can be determined using probabilistic design methods[18]. Studies carried out on a high-bay warehouse in Ebensee, Austria, have supplied additional, important findings on how different adhesives, the adhesive thickness, the length of overlap and the thicknesses of the parts being joined influence the stress distribution, plus potential methods of determining the stress distribution in the adhesive joint[19]. Owing to the agreement between the calculated and measured results, in future it will be possible to calculate the stress distribution in such adhesive joints very easily and check how various parameters influence the connection. These results open up new connection technology options for the future of engineered timber structures.

Besides creating nodes and junctions in a structure, lengthening beams and columns for building large structures is also important. Beam lengths are determined by production facilities and especially by transport options, where the economic limit to component lengths is 20–25m. However, members for large single-storey sheds and bridges are frequently much longer. The introduction of a new, patented, highly effective adhesive joint for timber beam splices enables glued laminated timber beam segments to be joined together on site by means of universal finger joints and a splice piece (also glulam) with two scarf joints[20]. High-strength timber laminations bonded across the butt joint in the tension zone strengthen and secure this joint yet further. This system allows practically any length of beam to be assembled from individual segments. With a degree of efficiency[21] of 0.9–1.0, the adhesive joint achieves the same strength as the timber beam itself[22].

The newly introduced technique of wood welding manages without any adhesives at all[23]. This jointing method exploits the thermoplastic properties of lignin in order to connect two timber components together. To do this, a rapidly vibrating piece of wood is pressed against a fixed piece so that frictional heat is generated at the surfaces in contact. At temperatures > 200°C the lignin begins to melt, liquefies and spreads out evenly over the entire surface of the joint. As soon as the vibrations stop, the wood cools and the lignin hardens under the action of the applied pressure. This adhesive-less connection, which can be used for discrete points or over larger areas, reaches its final strength after just a few seconds. As wood welding uses wood's own bond-

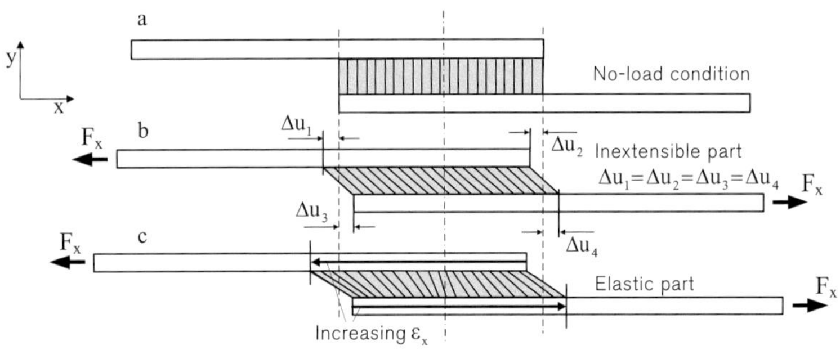

Lap joint in single shear in loaded and unloaded states. Joining together elastic parts leads to stress peaks at the ends of the overlap, which have to be accommodated by the adhesive.

ing forces[24], the strength of the welded joint differs for different species, varying between 5N/mm² for oak laminations and 10N/mm² for maple laminations.

As a follow-up to these investigations, the latest research concerns wood welding using powder-type adhesives and the development of adhesive-free connections for engineered timber structures.

Strengthening connections

The nodes and connections in long-span timber structures normally make use of mechanical joints with steel connectors. As the steel elements generally used reach only 65% of the tensile strength of the timber member, they usually represent the weak spots in a timber structure.

The load-carrying capacity of these connections is primarily influenced by the density of the timber or the bearing stresses at holes plus the dimensions, strength and arrangement of the fastening elements. Apart from that, the load-carrying capacity is further reduced by wood's anisotropy and the fact that the cross-section is weakened by holes, cutouts, etc. Although timber has a high strength parallel to the grain, it has only a low resistance to transverse tension, which is crucial around connections. Where steel dowels or nails are subjected to shear, high bearing stresses occur around

A 500m long footbridge in Anaklia, Georgia. The glued laminated timber beams up to 48m long are made up of smaller segments glued together on site, HESS TIMBER, 2012.

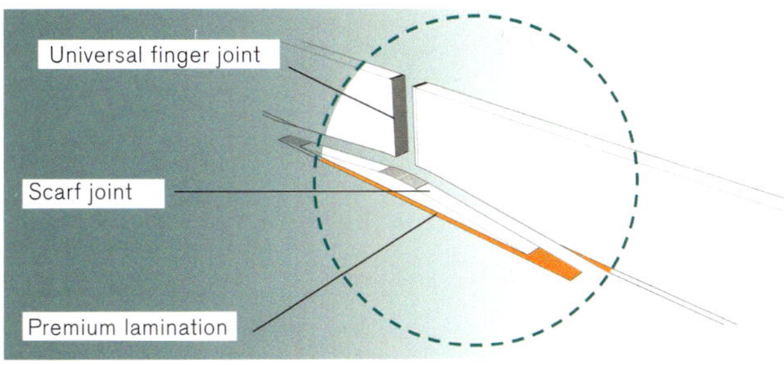

"HESS-LIMITLESS finger joint system".

18 See *Probabilistic design of adhesive debonded timber joints*, T. Tannert, S. Hehl, T. Vallée, in: *Bautechnik*, No.10, 2010 (in German).

19 See *Modellierung geklebter Stöße am Beispiel Hochregallager*, A. Thiel, in: Proc. of International Wood Construction Conference (IHF 2011).

20 The HESS company patented this joint as the "HESS-LIMITLESS finger joint system" in 2009.

21 The degree of efficiency of connections describes the load-carrying capacity of the joint (under the effects of flexural, normal and shear forces) in relation to the load-carrying capacity of the members themselves.

22 See *Geklebte Vollstöße großformatiger Brettschichtholzträger*, S. Aicher, in: IHF 2011, Vol. I.

23 Inspired by the automotive industry, which has been using linear vibration welding for many years to join together polymers or metals, Bern University of Applied Sciences in Biel, Switzerland, has been investigating this new wood jointing technique in collaboration with the Henri Poincaré University in Nancy, France, since 1993. The latter institute has also been researching rotary wood welding.

24 During the experiments, nuclear magnetic resonance analyses discovered new chemical bonds between lignin and furfural.

the holes in the timber, which leads to lateral buckling of the wood fibres and to lateral displacement of the fibres at the edges. The consequence is stresses perpendicular to the grain, which can cause shear or splitting failures. Connections are strengthened to achieve a balance between the load-carrying capacity of the connections and that of the structure itself. Up until now, failure due to splitting of the timber structure has been prevented by using threaded rods inserted and glued in place or by attaching wood-based products and punched metal plate fasteners on the outside. However, much more efficient and visually appealing is the use of self-drilling full-thread screws, which have now become established for resisting transverse tension around notches and beam connections, for resisting transverse compression at supports, for strengthening connections with steel dowels and for reinforcing timber beams[25]. Reinforcing highly stressed beam regions with full-thread screws inserted at an angle resists the shear stresses that occur and therefore allows the load-carrying capacity to be better utilised.

The use of textile fabrics and fibre-reinforced polymer composites in the form of patches – applied around drilled holes to reduce the bearing stresses and thus reinforce steel dowel connections – has been the subject of research for a number of years[26]. These patches consist of several layers of glass fibres embedded in a matrix. The glass fibres can either be provided as loops, spirals or star-shaped non-crimp fabrics to match the flow of forces around the dowel or as several plies of a unidirectional glass-fibre non-crimp fabric laid at different angles (0°/45°/90°). Combined with a thermosetting epoxy resin system, the result is a quasi-isotropic composite. Supplied as a semi-finished product in the form of 10mm thick circular patches with an outside diameter of 120mm and a hole diameter of 20mm, the patches are easy to glue into routed recesses in areas at risk of transverse tension.

When used to strengthen a bolted connection, the ultimate load is increased by more than 200% compared with an unreinforced connection. Increasing the efficiency of connections by strengthening them to resist transverse tension manifests itself in the form of longer spans and smaller member dimensions.

Upgrading the material

To improve the material properties of timber such as strength, stiffness and weather resistance (durability, UV and dimensional stability) and counteract wood's anisotropy, timber must be combined with other materials or its structure must be modified.

Techniques such as acetylation, waterproofing, heat treatment or impregnation with synthetic resins are all methods that have been developed and tried out in recent years with the aim of improving the weather resistance of wood and wood-based products. These environmentally friendly wood preservation methods have not been able to establish themselves so far because of their negative effects on the physical and mechanical properties of components, which are revealed in changes to the failure, bonding and loadbearing behaviour as well as a tendency to crack. However, these disadvantages can be offset if the techniques are employed in hybrid components in combination with untreated wood.

Using hardwoods, compressing the wood and producing moulded wood sections represent simple ways of increasing the efficiency of wood and hence the efficiency of timber structures. Producing glued laminated

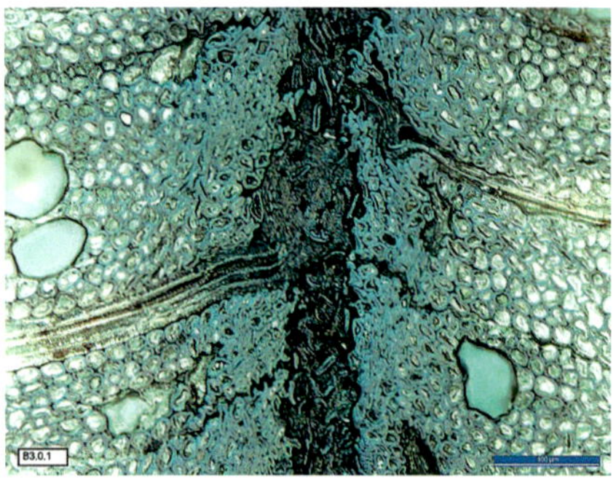

Fibres become detached from the cell structure and so a new network of fibres forms in the welded joint once it has hardened. It is this new network that is responsible for the strength of the connection in addition to the chemical bonds between the various constituents of the wood. The functionality of this environmentally friendly welded joint has already been tested on snowboards, which had a wooden core made from welded timber laminations.

timber from hardwood laminations, the simplest method for increasing the strength and stiffness of timber members, has failed so far because it has not been possible to produce a verifiable loadbearing and permanent bond between the laminations[27]. Whereas many adhesives are available for gluing softwoods, up until now no adhesive has been found that can bond hardwoods to match the requirements of a loadbearing joint. Research has established that urea-melamine-formaldehyde (UMF) resin adhesives can be used to achieve loadbearing glue joints between beech laminations, which has led to the granting of a national technical approval for the production of beech glulam members and beech hybrids[28] and has hence laid the foundation for the widespread use of this new material. The flexural, shear and bearing values of glued laminated timber members made from beech laminations are far higher than those of glulam members made from softwood; the flexural strength is almost double and the shear strength rises by a factor of 1.5. Notwithstanding, answers to the disadvantages in the manufacturing process, which is very time-consuming because of the relatively long closed time of 60 minutes and the 24 hour press time, plus the pronounced swelling and shrinkage behaviour of beech must be found in the future. Further developments in the adhesives and the thermal treatment could counteract these disadvantages.

Another possibility for increasing the strength, stiffness and durability of timber is the thermomechanical densification of solid timber to form so-called pressed wood. The production of this involves heating

25 *Selbstbohrende Schrauben und ihre Anwendungsmöglichkeiten*, H. J. Blaß, I. Bejtka, *Holzbaukalender*, Karlsruhe, 2004. *Verstärkung von Bauteilen aus Holz mit Vollgewindeschrauben*, I. Bejtka, *Karlsruhe Report on Timber Engineering*, Vol. 2, Karlsruhe, 2005.

26 *Hochleistungsholztragwerke – HHT – Entwicklung von hochbelastbaren Verbundbauweisen im Holzbau mit faserverstärkten Kunststoffen, technischen Textilen und Formpressholz*, P. Haller, M. Hamann, M. Hofmann, TU Dresden.

27 The production of glulam from hardwood laminations has been the subject of research since the 1960s (Egner & Kolb, 1966; Gehri, 1985). However, the problem of gluing for structural engineering applications has never been solved. Despite the lack of standards, in Switzerland, individual parts of loadbearing structures have been made from glued hardwood laminations (beech and ash) over the last 25 years, e.g., Dörfli Bridge, Eggiwil (1985), but each application requires individual approval by the authorities.

28 The research project was carried out by TU Munich (Holzforschung München) in conjunction with Karlsruhe Institute of Technology (KIT) and partners from industry (Schaffitzel, Obermeier, Türmerleim). October 2009 saw the granting of a national technical approval for the production of beech glulam up to strength class GL 48. As a pilot project, the extension to the Bavarian State Institute of Forestry in Freising is to be built in beech glulam.

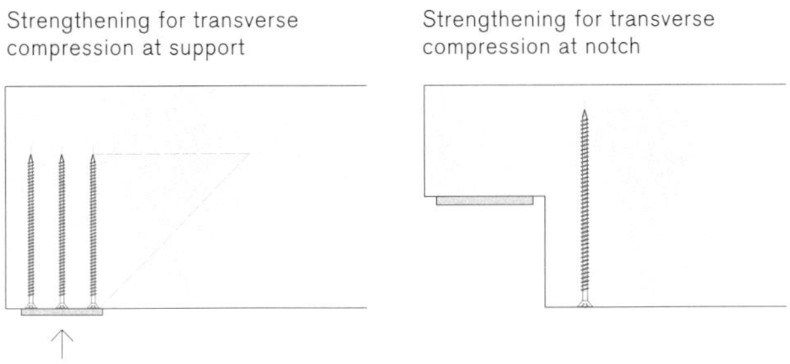

Examples of members reinforced with full-thread screws.

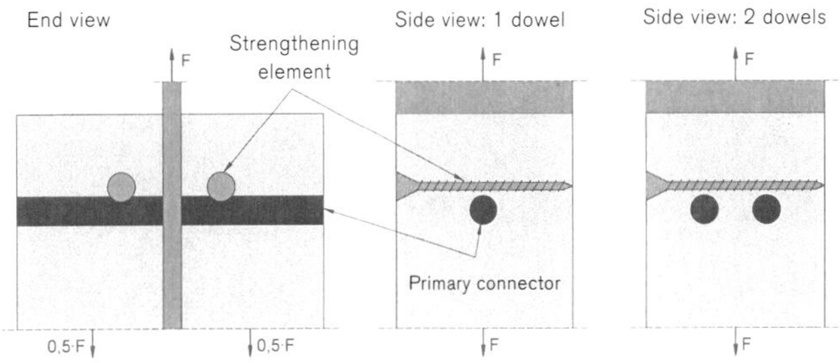

To strengthen steel dowel connections, the screws are inserted directly alongside the steel dowel at 90° to the grain and at 90° to the axis of the bar-type connection. This prevents the wood splitting and generates a support effect for the connecting elements, which in turn reduces the displacement and almost doubles the load-carrying capacity of the connection.

the solid timber to about 130°C with heated moulding platens (microwaves or convection, with steam to conduct the heat) and compressing it in a press at 5MPa perpendicular to the grain. As the pores account for about 60% of the volume in softwood, the cross-section can be reduced by about half.

Using a degree of compression of 50% doubles the tensile and compressive strengths parallel with the grain, increases the flexural and shear strengths by factors of 2.5 and 1.7 respectively and improves the compressive strength perpendicular to the grain by a factor of 4.5. Further heating beyond 200°C increases the biological resistance, which in turn leads to better weather resistance. Similarly to a number of hardwoods, however, these improvements to the material properties are offset by a number of disadvantages when it comes to processing the material. Owing to its high density, pressed wood tends to split and adhesive joints are difficult to form.

The production of moulded wood sections represents another attempt to improve the efficiency of the material in the timber cross-section[29]. The disadvantages of timber structures built exclusively from solid timber sections could be overcome by producing efficient timber sections with different cross-sectional forms in a similar way to the standardised steel or plastic sections manufactured by industry. As an example, in the laboratory, circular hollow timber cross-sections made from squared sections with 30% compression were glued together with resorcinol resin to form boards. Heat and moisture are used in the moulding process and so the cells of the pressed wood "unfurl" and the individual pressed wood cross-sections adapt to the curvature of the wooden tube. The circular hol-

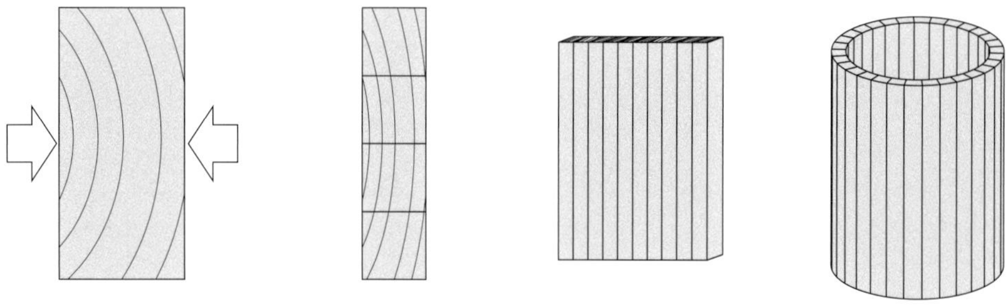

Production of moulded wood sections from compressed squared sections. The ultimate load of a textile-reinforced circular hollow timber column with an outside diameter of 274mm and a material thickness of 19mm corresponds to that of a solid round timber column with a diameter of 180mm.

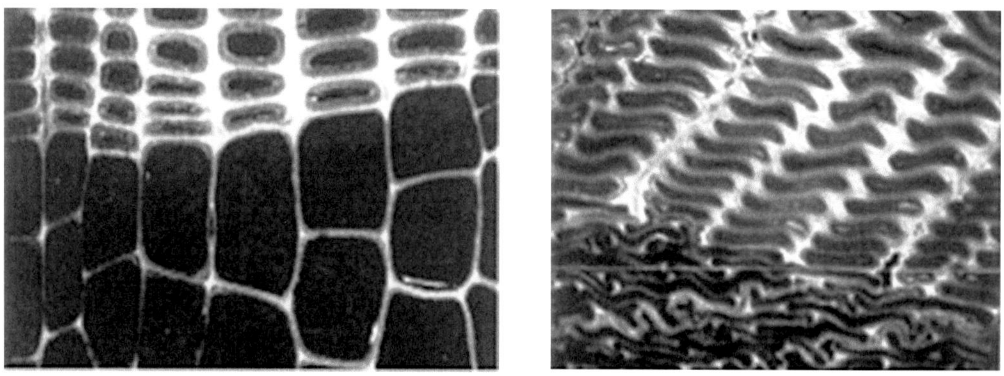

Compressing solid timber causes the cell walls to collapse and the pores to close. This change in the cell structure essentially results in a new material with new properties that vary with the degree of compression.

low timber sections can be joined together with finger joints to produce members of any length.

Hybrid components

One specific way of upgrading the material is to produce hybrid components. Hybrid components consist of a combination of different materials or substances in which the positive material properties of the respective parts play a role. The timber hybrids produced and tested to date involve conventional glued laminated timber with laminations made from timber of higher strength grades, laminated veneer lumber (LVL), thermally modified timber (TMT), compressed laminated wood, fibre-reinforced polymers, steel strips, steel reinforcement or textile fabrics. In order that the combination is effective and the different properties of the materials and substances can be used according to the respective specification for the component, the quality of the glue joint is crucial.

The adhesive bond is influenced by the surface properties (smooth, rough, open pores, closed pores), the microclimatic relationships at the boundaries between the glue and the parts being joined and the deformations of those parts, which are often different and are caused by loads, shrinkage, swelling and creep. The nature and duration of the loading is just as critical for the long-term serviceability of the bond as is the behaviour of the adhesive under different climatic influences such as heat, cold, solar radiation, moisture and dryness. In addition, the processing (one- or multi-part adhesives, thickness of adhesive, type and duration of curing, etc.) and aspects such as gap-filling properties or the nature of the bond itself (rigid, elastic, plastic) are vital when selecting an adhesive. A suitable

29 A comprehensive, large-scale research project at TU Dresden, in cooperation with HESS TIMBER GmbH, is looking into different options for establishing wood as a high-performance material. Moulded wood sections, with their efficient material properties, have been developed and tested as part of this project. *Hochleistungsholztragwerke – HHT – Entwicklung von hochbelastbaren Verbundbauweisen im Holzbau mit faserverstärkten Kunststoffen, technischen Textilien und Formpressholz*, Dresden, 2011.

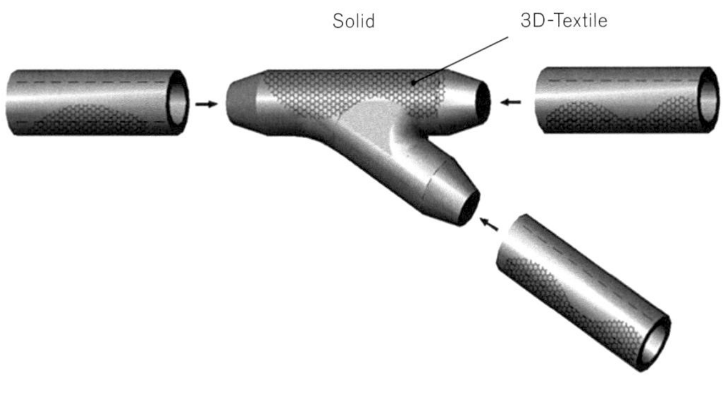

The hollow sections can be used to build forks for tree-like columns. Spigots about 20cm long allow the round moulded wood sections to be fitted onto the fork, to which they are glued to form tree-like branches.

adhesive must be found or developed to suit these parameters and the respective hybrid structure. New hybrid structures are mostly tested with known adhesive systems, which are then modified accordingly to suit the requirements. Owing to the numerous parameters that influence the bond system, the development of hybrids is very complicated, so it is no surprise that so far only a few systems have been used in practice.

Whereas hybrid beams made from softwood in combination with beech laminations top and bottom have been granted national technical approvals and therefore can be readily used for long-span structures in the future, the material properties of other hardwoods such as poplar, oak, ash and ailanthus are currently still undergoing testing to establish whether they can be combined with softwood for loadbearing purposes in engineered timber structures. As the flexural strength of such beams is essentially determined by the tensile strength of the laminations top and bottom, beech hybrid beams reach similar load-carrying capacities and stiffnesses as glulam beams made exclusively from beech[30] and therefore represent a much less costly alternative. Another effective method is partial replacement of the top and bottom laminations in order to strengthen conventional glued laminated timber at heavily loaded supports and junctions. This raises the degree of efficiency of the connections to 100%, which in turn has a positive effect on the load-carrying capacity of the entire structure. In Switzerland, ash, despite the lack of standards for design methods, has been used in a few cases for local strengthening at supports and even, just recently, for building complete structures from ash glulam or ash hybrids[31].

The ash glulam beams (GL 48) for a new sports hall in Sargans, Switzerland, are additionally reinforced with steel and therefore reach strength grade GL 60. The load-carrying capacity is therefore more than double that of a conventional glued laminated timber beam; combining with concrete reduced the dimensions of the beam even further. The appearance of these extremely slender timber sections, 140 × 500mm spanning 10.65m, is very different from that of a conventional timber structure.

The idea of increasing the efficiency of glued laminated timber beams by replacing the top and bottom laminations is not new; tests on hybrid beams with

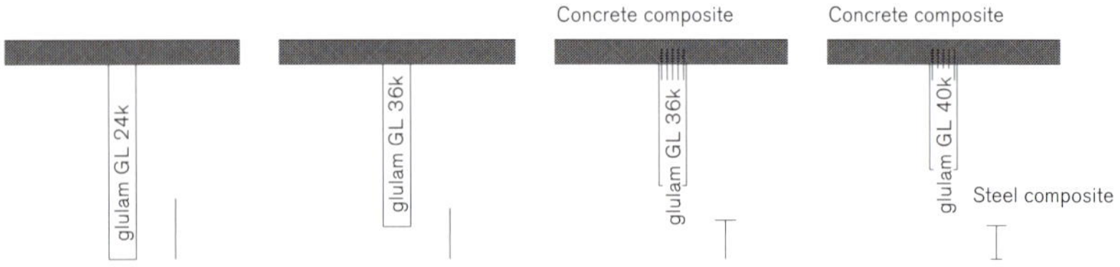

The high shear stresses near the supports were critical for the use of glued laminated timber beams with ash laminations. Ash glulam reaches a shear strength 1.5 times that of conventional glued laminated timber. Large sports hall in Sargans, Neue Holzbau AG, Lungern, Switzerland, blue architects and Rubrecht Architekten, 2012.

laminated veneer lumber (LVL) in the tension zone were carried out as long ago as the late 1970s[32], and the hybrid cross-section was proved to have a higher load-carrying capacity. As a continuation of this research, hybrid structures with compressed laminated wood have been tested in recent years[33]. Compressed laminated wood consists of wood veneers 1.5–4.0mm thick which are pressed under the action of heat, impregnated by immersing in phenolic resin and glued together. Compressed LVL excels through its very high strength and stiffness plus its good weather resistance, which results in durable and dimensionally stable structures[34]. However, owing to its very unfavourable creep behaviour caused by the high proportion of resin, the material may not be used for loadbearing members. Combined with glued laminated timber, the respective weaknesses of the materials can be offset. Loading, delamination and weathering tests on hybrid beams made from larch glulam with top and bottom laminations of compressed laminated wood show positive results: the load-carrying capacity could be increased two-fold, the stiffness by a factor of about 1.5 and both the glue joint and the wood could cope with the exposure to the weather. The long-term behaviour of the hybrid beam is currently being tested in a pilot project[35].

With their low weight, extremely high tensile strength and easy workability, fibre-reinforced polymers are ideal for combining with wood. Accordingly, research in this field has been going on since the mid-1990s. The tests initially involved timber combined with polymers reinforced with glass, aramid and carbon fibres. However, after the turn of the century, industrially manufactured CFRP[36] (carbon fibre-reinforced polymer) strips, well known in reinforced concrete, became firmly established. Owing to the high price of carbon fibres, in the early years the polymer laminations were used for specific, partial strengthening in areas with high tensile or shear stresses or for strengthening existing floor beams in refurbishment and conversion projects. To do this, 50mm wide polymer strips were inserted into slots cut in both sides of the beam in the tension zone and glued in place with a mixture of two-part epoxy resin and quartz sand. The gap-filling properties and the possibility of being able to produce a glue joint without using pressure makes epoxy resin perfect for this type of application. In poly-

30 *Biegefestigkeit von Brettschichtholz-Hybridträgern mit Randlamellen aus Buchenholz und Kernlamellen aus Nadelholz*, M. Frese, H. J. Blaß, Karlsruhe, 2006.

31 Ash glulam was used for parts of the roof to a skiing facility in Arosa and ash laminations were used to strengthen supports at a warehouse in Conthey.

32 *Bending strength of small glulam beams with a laminated-veneer tension lamination*, M. O. Braun, R. C. Moody, *Forest Products Journal* 27, 1977.

33 *Hochleistungsholztragwerke – HHT – Entwicklung von hochbelastbaren Verbundbauweisen im Holzbau mit faserverstärkten Kunststoffen, technischen Textilen und Formpressholz*, P. Haller, M. Hamann, M. Hofmann, TU Dresden.

34 Resin-treated laminated compressed wood (Compreg) products have long since been used in electrical engineering. Their use in architecture and structural engineering is relatively new.

35 The hybrid beam made from glulam strengthened with compressed laminated wood was patented in 2009. It is to be tested in a bridge structure as part of the HHT research project at TU Dresden.

36 CFRP strips have been used for upgrading and repairing reinforced concrete slabs since the early 1990s. Strips developed in the USA have been used to strengthen timber beams in Switzerland and the USA as "FiRP® technology" (fibre-reinforced plastic) since the end of the 1990s. The first timber structure in Europe to be reinforced with aramid fibres was the warehouse of the Bürli company in St. Erhard, Switzerland (1999, Zöllig Holzleimbau AG).

mer/timber hybrid beams for use in engineered timber structures, the polymer laminations form interlayers in the tension zone of the glued laminated timber[37]. In order to increase the efficiency of the laminations in a hybrid timber beam, the use of prestressed fibre-reinforced polymer laminations is currently being tested for practical applications.

Textile/wood hybrids represent another new development[38]. Inspired by boatbuilding and wind turbines, which use textile reinforcement to protect timber rotor blades against moisture and improve dimensional stability, TU Dresden has developed textile-reinforced timber columns.

Combination with industrially manufactured fibres and textiles compensates for the low transverse tensile and shear strengths of timber components, improves the bearing stress values at holes and protects the timber against the weather. Buckling tests on round moulded wood sections made from thermally compressed spruce combined with hose-type textile fabrics, which were applied with thermosetting resin systems, increased the ultimate load by a factor of 1.6 compared with the unreinforced specimen, and also resulted in a ductile failure behaviour. In textile-reinforced hollow box sections made from glued spruce with a ±45° fibre alignment, the compression loads were as much as doubled. Crucial for the degree of efficiency of textile reinforcement is the angle of the fibres, the degree of strengthening in terms of the number of layers of non-crimp fabrics or proportion of fibres and the choice of

37 The reader is referred to the research work carried out at the University of Karlsruhe, Chair of Timber Engineering & Building Design, H. J. Blaß, M. Romani & M. Schmid, or Daniel Tingley's presentation at WTC in Montreau, 1998. In the meantime, the Buchacher company in Austria has built a number of timber structures with polymer/timber hybrid beams. Trade fair hall 2 in Klagenfurt and the timber structures on the premises of furniture producer Blaha are strengthened with glass fibre laminations.

38 See the HHT research project, TU Dresden.

39 The research project was carried out by Holzforschung Austria in collaboration with the science and technology universities in Vienna, Munich and Lausanne. Researchers have been investigating timber-glass composite elements since the late 1990s; see John Pye (University of Bath), Peter Niedermaier (TU Munich) and Wolfgang Winter (TU Vienna).

40 Float glass, heat-strengthened glass and laminated safety glass are all suitable for composite elements. The use of a timber coupling strip to join glass and timber was patented in 2005 as a result of the research project. A GFRP coupling strip is now manufactured by Knapp GmbH.

Combining polymer laminations with glued laminated timber increases the load-carrying capacity of a beam by 70% and the flexural stiffness by 30% compared with unreinforced glulam beams.

material. The different parameters can be adapted to the specific structural requirements and therefore increase the efficiency of the strengthening.

Composite structures

Besides the development of hybrid components, recent years have also seen a multitude of research projects involving timber composite designs. The outcome up to now is timber-concrete composite members for bridges or suspended floor slabs. A combination that exploits the tensile strength of timber and the compressive strength of concrete results in economic long-span designs that also have good acoustic and fire protection characteristics.

Loadbearing timber-glass composite elements represent the latest development in this field[39]. The high compressive strength of glass makes it suitable for carrying the horizontal or vertical forces that occur in façade and beam designs. In a façade design patented very recently, panes of glass glued in place can be used as bracing elements in buildings with one or two storeys[40].

This form of composite element only became possible after the introduction of elastic adhesives into timber construction. Owing to the different material properties of glass and timber, a rigid glue joint would lead to stresses in the system which in turn could lead to cracking, even failure. On the one hand, the glue joint must be resilient in order to accommodate stresses and

Moulded wood tubes with different types of textile reinforcement (carbon, glass and aramid fibres) have been tested at TU Dresden; Prof. Peer Haller, Chair of Timber Construction and Structural Design.

at the same time transfer the forces from one element to the next. On the other, the loads must be transferred carefully and evenly so that the pane of glass can function as a bracing element for the building. Elastic glue joints, common in shipbuilding and the automotive industry, satisfy this requirement. A timber tower built in 2003 served as the first pilot project[41]. The stiffening glass panes made from laminated safety glass, which were connected to the posts and rails elastically by a coupling strip, were positioned behind the actual façade. There was still a lack of knowledge about the properties of elastic adhesives and the requirements resulting from the composite element as well as suitable methods for analysing the structural behaviour and designing the structural elements, which meant that the system could not be used as a loadbearing façade. The particular features of an elastic adhesive joint (instead of requiring clamping and press time, the adhesive bond is achieved from the surface structure of the parts to be joined and the application of the adhesive) had to be researched. Likewise, how to determine the influence of the geometry of the layer of adhesive (governed by the overlapping surfaces, the thickness of the adhesive and the configuration of the cross-section at the glue joint) and how the nature and duration of the loads affect the strength of the composite element also needed to be investigated. The findings and the newly developed method of calculation (lattice spring model for predicting deformations and shear field method for determining load-carrying capacity) enabled the construction of the first building with a timber/glass composite façade in 2011[42].

Another area of research is opening up composite construction with timber and polymer sheets. A bridge structure in Darmstadt has supplied practical experience of this[43]. The approx. 3m deep main bridge beams are composite members with glued laminated timber beams as top and bottom flanges and a 70mm thick PMMA (polymethyl methacrylate) sheet as the web. The PMMA sheet is clamped between the two-part timber flanges (2 No. 150 × 200mm) top and bottom and connected to these with M20 close-tolerance bolts every 42cm. The comparatively low modulus of elasticity of PMMA is an advantage for this composite construction: prior to the sheet reaching the limits of its loadbearing capacity, the structure fails due to the more brittle timber members. A disadvantage, on the other hand, is the high coefficient of thermal expansion of the thermoplastic material: the stresses that occur in the system at high temperatures must be compensated for in the details and the connections.

Many of the new developments described above are already changing the face of timber construction permanently. At the same time, the research work illustrates the potential for timber structures. Whereas the introduction of new, high-performance connections, materials and hybrids increases the efficiency of timber structures and alters their appearance, combining several measures could result in an exponential increase in efficiency. The reinforced hybrid beams made from ash glulam in grade GL 60[44] or the glued, scarf-jointed timber connections for long-span timber structures are good examples of the future for building with wood.

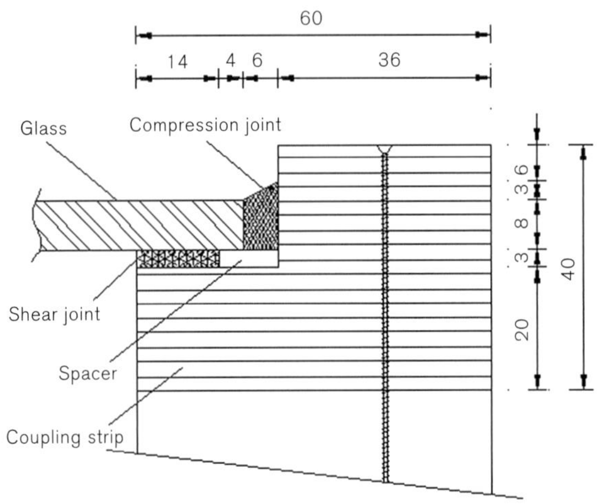

The pane of glass is bonded elastically to a coupling strip under controlled factory conditions and then fitted into the timber structure on site.

The timber/glass composite elements in the façade brace the whole building. New research is concerned with expanding the range of applications and improving the system; in future the silicone adhesive used is to be replaced by more efficient acrylic or polyurethane products. Using such composite elements in multi-storey buildings or as timber-glass composite beams represent further potential applications. "Schattenbox", Superlab, 2009, www.superlab.at.

Timber-polymer composite beams. The composite beams, 26m long in total, are designed as single-span members with overhanging ends. The grade GL 28h glulam chords resist the tensile and compressive forces and the PMMA sheet resists the shear. Bridge over former palace moat, Darmstadt, Germany, 2008.

41 The tower was 11.20m high and square on plan (3.60 × 3.60m); it was originally erected at the BAU trade fair in Munich in 2003 to demonstrate different types of glass. After the fair, the tower was set up on the company's premises.

42 The house in Eichgraben, Austria, was designed and built by the architects Dold and Hasenauer working together with Holzforschung Austria.

43 The composite construction was developed by TU Darmstadt in conjunction with plastics manufacturer Evonik Röhm GmbH; the footbridge over the former moat to the palace in Darmstadt was built in 2008 as a pilot project.

44 See "Neue Holzbau", *in: Holzforschung Schweiz*, 01/2011.

Developments in timber construction materials

Khaled Saleh Pascha

Modern methods of production and new wood-based materials have had a very significant effect on the trend towards long-span, efficient loadbearing structures in timber. As already pointed out in the chapter on historical developments, long-span timber structures, going beyond the dimensional limits set by the tree trunks themselves, could be built only with considerable effort.

Glued laminated timber

Building with glued laminated timber (glulam for short) revolutionised timber construction in the 20th century. After its introduction at the start of the century, it was possible to erect structures that in the old days of timber construction would have been very difficult, i.e., the fabrication of timber members with dimensions greater than those of the original material, the tree trunk. The gluing of individual laminations in a continual process meant that, theoretically, there would be no limits to the size of loadbearing elements that could be produced. Furthermore, it was now possible to produce curved beams or columns in timber, following the ideal line of the bending moment diagram, without any significant wastage.

Glued laminated timber can be seen as the technological successor to laminated plank construction (Emy form of construction). The new form of construction prevented the buckling of the individual laminations under bending actions, which was a problem in laminated construction. Gluing together individual 30–40mm thick, planed timber laminations to form continuous timber members creates a homogeneous composite material that overcomes the length and width limitations of sawn timber. In principle, this jointing technology allows the production of timber members without any limits to size or form. Another great advantage of glued laminated timber is that weaknesses in the timber microstructure, e.g., heartwood, knots, resin accumulations or areas of bark, are cut out accordingly in order to arrive at a guaranteed grade of timber. This overcomes the problem of sawn timber, where even just one small knot can weaken the structural quality of the entire member quite decisively.

Gluing together pieces of timber to create large, coherent timber members must be seen as the prerequisite for longer spans in timber construction because it had been impossible to "tame" the deformation behaviour of the timber before this technology appeared.

The patents of master carpenter Karl Friedrich Otto Hetzer dating from 1891–1910 were crucial here. They concerned certain timber cross-sections (flooring system, timber beam) assembled from smaller pieces to form a structural unit. The patent dating from 1906 (DRP No.197773) was for a curved timber component,

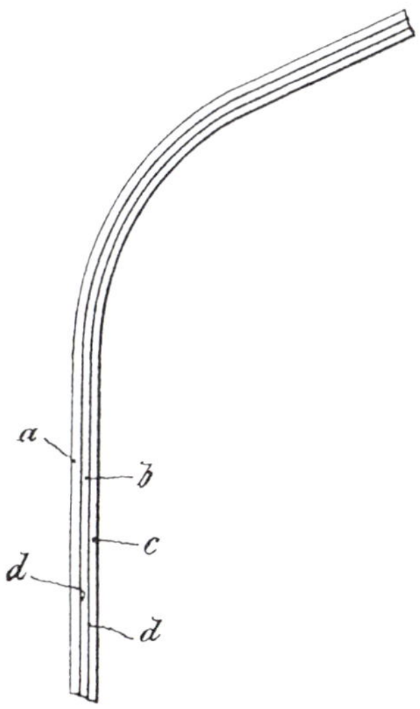

Otto Hetzer's patent No. 197773 of 1906 for a curved timber member that is both column and rafter at the same time.

Auditorium during construction. The loadbearing structure to the large auditorium would have been economically and technically impossible without the options of glulam, which allows the production of curved timber elements with very little wastage. Centro Cultural Matucana 100, Santiago de Chile, Chile; architect: Martín Hurtado, 2002.

a combined column and rafter made from several laminations, which was essentially a glued laminated timber beam.

This patent in particular can be regarded as a milestone in modern glued laminated timber construction. The combined column/rafter component described in Hetzer's patent demonstrates an especially typical and ideal use among the many possibilities for this new approach to producing curving forms in timber without wastage. The individual timber laminations were bent to the desired shape and then glued together. The gluing "locks" the separate segments together, that is, the curved form of the component is retained and it takes no more wood than that actually needed to create the form. Gluing overcame the restriction of only being able to use straight elements in timber construction. Wooden planks could now be produced bent to a radius that stressed them to the limits of their capacity and then glued to others, which fixed the shape once the adhesive had cured. Depending on the structural system, the prestress introduced into the wood fibres due to the bending of the pieces of timber can result in a further improvement to the structural response.

Absolutely crucial to the durability of the joints between the individual laminations is the adhesive technology available. In this context, it is important to realise that gluing is not a traditional method of joining wood, but instead is very closely linked with the evolution of modern engineered timber structures. In contrast to mechanical fasteners, which create a discrete, point-like connection and are often not rigid, a glue joint provides a rigid, two-dimensional connection, which is a major advantage for developing homogeneous components assembled from smaller segments[1].

Even though gluing techniques were known in the ancient world, up until the modern age this form of jointing had been used less in construction than in the building of furniture or musical instruments. The customary bone glues (animal glues) of those times required considerable work and were associated with high costs – factors that could not be justified in the building industry. In addition, there were the structural and constructional limitations due to the low quality of the glue and its inadequate weather resistance.

At the time of the invention of glued laminated timber in the early 20th century, the most common glues used for joints in the building industry were casein

1 See J. L. Moro, B. Alihodzic, M. Weißbach: *Baukonstruktion – Vom Prinzip zum Detail.* 3: "Umsetzung", Berlin 2008, p. 260.

The natural limits to the sizes of structures made from solid timber can be overcome by using glulam. The curving bottom chord of the truss would have been much more complicated, expensive, wasteful and less efficient structurally if it had not been made from smaller pieces glued together to form a curve. Glued laminated timber trusses at WIEHAG, Altheim, Austria, 2008.

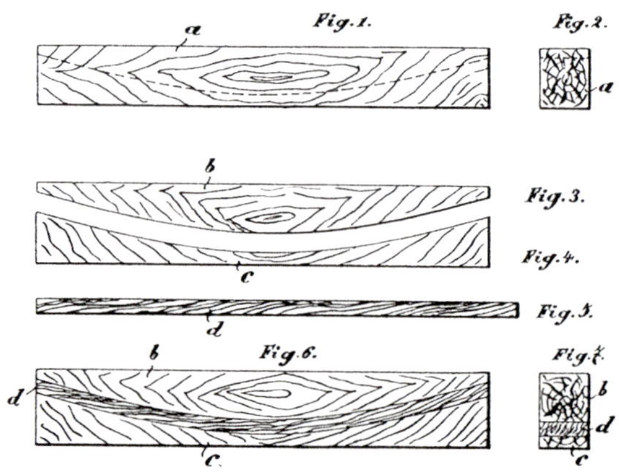

Beam made from glued pieces of timber, c. 1910.

The salt warehouse belonging to Salinen Austria AG in Ebensee, Austria, is an interesting example of the use of glued laminated timber and demonstrates the leap in dimensions made possible by this technology. With a length of 110m, width of 22m and height of 21m, this building can be counted among the largest exclusively timber structures. The fact that not only are all the horizontal and vertical loadbearing elements in glulam, but all the connections between the individual elements are traditional dovetail joints (cut on CNC plant), is the outcome of the particular function of the building as a warehouse for salt. In this case the very high corrosive load due to the salt rendered steel – otherwise so prevalent among high-bay warehouses – unsuitable. The cost of multiple coating of the steel structure and the extra maintenance would have led to an unjustifiable economic solution, whereas a timber structure requires no additional treatment to protect it against salt. The 96cm wide × 12cm thick plate-like columns are fully prefabricated as high-bay racking modules in one piece over their full height of 21m. Wind girders of steel at the end of the building, or between the columns, ensure the necessary stability. In this project, the great advantage of glulam was not the design freedom of being able to create curved forms with comparatively little effort and wastage of material, but rather the possibility of fabricating timber elements with very large dimensions in one piece. The structural and erection advantages of 21m high continuous columns outweighed the extra cost of transporting such large members.

High-bay warehouse belonging to Salinen Austria AG in Ebensee near Salzburg, Austria; design and construction: Kaufmann Bausysteme GmbH, Reuthe, Austria.

glues made from naturally soured casein of milk, which was mixed with calcium compounds. These traditional, common, casein glues were hardly waterproof (and the animal glues based on organic protein-phosphorus compounds not at all) and had only low shear strengths, which frequently resulted in joints coming apart as a result of inherent stresses in the wood. Long setting times affected both types of glue. So for a long time glue was the weak link that prevented truly homogeneous composite action between wood and binder. Once the first modern, cold-curing synthetic resin glue, based on urea, appeared in 1928 ("Kaurit"), the construction industry had at its disposal a jointing substance that was fully waterproof and resisted fungal growth[2]. In the course of their further development, these modern synthetic resin glues underwent improvements and these days achieve a higher adhesive bond in the glue joint than between the wood fibres themselves, i.e., failures of glued timber components generally occur in the wood and not in the adhesive. Nowadays, modern synthetic resin glues are inexpensive to produce, very easy to use and, depending on the type of glue, allow their setting times to be adjusted to the requirements. These aspects together with their weather and ageing resistance add up to a triumph for modern glued laminated timber construction which would have been inconceivable without the availability of appropriate gluing technology.

Another step directly associated with the development of modern glues revolutionised the production of

Switzerland's central salt depot in Rheinfelden employs a totally different structural concept for the storage building. In contrast to the high-bay warehouse in Ebensee, the de-icing salt stored here is protected from the weather by a dome 93m in diameter and 31m high. This is the largest dome in Switzerland and the building can store 80 000t of salt. Some 1500m³ of indigenous silver fir and spruce, sourced locally, were used to build the salt depot in Riburg, the largest such facility built in Switzerland to date. The salt content of the air in the building has a preservative effect on the wood, allowing it to be installed untreated. The dome structure consists of a lattice of 402 glued laminated timber ribs bolted together at the intersections by steel nodes, Ensphere© connectors. This patented form of assembly was developed by Häring, the timber contractor responsible for this project, who had already used this principle to build a whole series of similar timber lattice shells for storing loose materials under the name Saldome®.

Salt warehouse for Schweizer Rheinsalinen in Rheinfelden, Switzerland; design and construction: Häring & Co. AG, Pratteln, Switzerland, 2004–2005.

timber beams and columns with large dimensions: the finger joint.

The most logical method for producing columns and beams longer than the natural limits to the dimensions of tree trunks would be to glue timber cross-sections together. However, gluing perpendicular to the grain of the wood is impossible and so the finger joint was developed. The wedges cut in the longitudinal direction of the wood provide a sufficiently large contact area approximately parallel with the direction of growth of the wood, which allows pieces to be glued together. Automatic finger-jointing machines cut the finger joints, up to 50mm deep, in the faces to be connected and then glue these together under pressure.

In glued laminated timber, the individual pieces of wood are joined with finger joints to form continuous laminations, which means that the joints between the pieces are no longer potential weak spots within the member as a whole. This together with the multi-ply lay-up, high-strength, durable adhesives and finger-jointed butt joints enables glued laminated timber construction to achieve much higher load-carrying capacities than solid timber of the same cross-section. And what is especially important for engineered timber structures is that by excluding imperfections and homogenising the quality of the timber in the compound component, it has been possible to standardise timber qualities to a large extent and hence achieve a structural predictability – the prerequisite for developing timber structures with challenging structural engineering aspects.

One aspect fundamental to the rise in quality, and hence the efficiency, of glued laminated timber in general is the quality assurance achieved by using specific, verifiable and standardisable timber grades. The growth-related imperfections frequently found in solid timber e.g., material weaknesses due to knots, resin accumulations, rotted parts, etc. can be detected and removed. The remaining segments, corresponding to the specified quality requirements, are joined together with finger joints and therefore create a continuous lamination with a consistent level of quality which can be processed to form a glulam element in subsequent operations.

It is not only the removal of knots or areas with distorted growth which permits a uniform classification of the wood. Further possibilities for prior grading of the timber laminations by means of automated quality control systems using x-rays, scanning and mechanical tests (bending) enable classification and grading with respect to hidden properties according to the structural requirements of the component. Consequently, correspondingly high-value timber laminations graded according to quality can be selected for heavily stressed components. It is also possible to mix given grades so that the qualities of individual timber laminations within a cross-section can be employed optimally and efficiently to suit the structural requirements: the laminations in the middle of a beam cross-section are less highly stressed and therefore can be made from lower-grade laminations, whereas the outer laminations in the tension and compression zones will have to be made from higher-grade material. This leads to better utilisation of the materials used and, in the end, to lower production costs. Furthermore, such prior grading of the timber laminations also amounts to a visual quality control because the outer laminations of the timber member can be made from a material with fewer knots and a better appearance than the concealed inner laminations. And finally, it is possible to produce solid timber elements with an almost ideally adapted material quality and with a material homogeneity that achieves a load-carrying capacity about 80% higher than that of normal sawn or solid timber.

Yet another advantage of glued laminated timber is its better splitting resistance compared to solid timber. This can be attributed to the gluing together of three or more laminations and the associated homogenisation of the total timber cross-section.

Building with glued laminated timber renders possible a whole range of designs. Gluing together curved laminations in appropriately shaped dies enables the

2 See W. Rug: "Innovationen im Holzbau – Die Hetzerbauweise" (Teil 2), in: *Bautechnik 72* (1995), Vol. 4, p. 236.

production of a large number of different geometries for columns and beams. With the help of dies and moulds, individual laminations can be bent to a radius and glued under tension to create large elements while retaining the longitudinal direction of the grain of the wood. That would be impossible in conventional solid timber, where curved elements would have to be cut out of solid sections, resulting in high wastage.

It is this latter feature of glued laminated timber in particular that had a decisive influence on the design language of timber construction in the 20th century. Curving columns and beams plus resolved, seemingly interwoven grillages enrich the architectural vocabulary of contemporary timber construction. In a similar way to the appearance of modern formwork technology

The arches to the trade fair hall at Wels, Austria, designed by AT4 Architekten and finished in 2007. With a length of 90m, width of 6m and height of 12m, these arches spanning over the 14 400m² of uninterrupted floor space in this exhibition hall represent an impressive example of the possibilities of glulam construction. The outer layer in the opening between these pairs of splayed arches is a "wearing course" made from vertically laminated glulam members which, exposed to the weather, can be replaced at a later date. The direction of the grain alone reveals that these layers are non-loadbearing.

One example of the artistic and technical possibilities of glulam for creating non-linear elements with dies is the 143m long × 11m high building for wine production at the Perez Cruz vineyard in Maipo Valley in Chile. It was designed by Chilean architect José Cruz Ovalle. The curving glulam columns supporting the roof with its long overhangs continue as ribs inside the building and at the same time form the structure for a timber barrel vault. At the crown of the vault, protected by the overhanging eaves of the flat roof constructed clear of the barrel vault, there are large openings that admit daylight to illuminate the production and office zones.

Although the curving columns outside do not correspond to the optimum structural form for the bending moments and the associated deformation under vertical loads, timber is easily able to handle these bending stresses over a long period without suffering any damage. Notwithstanding, the architectural design options of these curved beams and columns are impressive. Such curving beams, up to 10m long and with a rise of 1m, can only be produced economically in glued laminated timber.

Pérez Cruz vineyard in Chile; architect: José Cruz Ovalle, 2000–2002.

The same type of project, a production facility for wine, this time in northern Spain, has very similar, albeit reduced, architectural language and, in particular, the same glulam technology for constructing the parabolic arches of the primary structure. Those are the features of this 18m span timber structure near Valladolid designed by Rogers Stirk Harbour. The primary structure makes use of three-pin arches, a statically determinate system that requires only hinged connections at mid-span (the crown of the arch) and at its supports. This type of structure is characterised by an especially high material efficiency and, due to the pinned supports, simple foundations, which results in effective, economic designs. The large arch spacing of 9m means that correspondingly deep and sturdy secondary beams are necessary plus tertiary members parallel with the main arches. Owing to the V-shaped steel members between the main arches and the secondary beams, the roof itself seems to float above the arches. These splayed steel struts increase in length towards the ground, allowing the roof shells to meet at a specific height and create the impression of one uninterrupted production area. An air space beneath the terracotta panels forming the roof covering, the deep basement in concrete and the 9m deep steps in the end façades prevent an excessive build-up of heat inside the building in summer and ensure an even temperature throughout the building over the whole year. An even temperature is one of the basic requirements for the production of wine and, above all, its storage (in the basement).
As with the vineyard in Chile, the use of glulam technology on this project is limited to the production of elements in single curvature, which does not require any complicated, computer-assisted production. Simple dies are sufficient here. The laminations are fitted up against the dies under tension and glued together to achieve the desired shape.

Bodegas Proto, Peñafiel, Valladolid, Spain; architects: Rogers Stirk Harbour + Partners; structural engineers: Arup/Boma/Agroindus, 2003–2008.

This leisure centre is located in France not far from Geneva across the border in Switzerland. It includes a shopping centre, a hotel, restaurants, sports facilities and a large water park. The shopping centre and sports facilities, a total area of about 15 000m², are housed in a gently curving, narrow structure with a reinforced concrete frame and green roof, whereas the covered part of the water park, with its undulating roof supported by a timber space truss, stands out distinctly against the Alpine panorama in the background. The arched timber structure over the water park, 120m long × 65m wide, covers a total area of 4300m². The roof structure consists of 14 two-pin arches spanning between 30 and 42m. These trussed arches with a maximum depth of 2.97m have curving top and bottom glulam chords made from indigenous fir which are 320mm in diameter. A total of 1300 glued laminated timber elements constitute the secondary structure. These are also round members which complete the space truss in the longitudinal direction. They connect the trussed arches together in such a way that a type of three-chord truss is formed, defined by the two trussed arches and one top chord of curved members located in the middle of the bay. The patented Résix® system from SAS Simonin is used for the connections between the individual timber members. This system uses metal rods glued with epoxy resin into holes drilled in the timber to join the members together. Being embedded in the timber, the metal parts are protected against corrosion in the chloride-laden air of the swimming pool.

The secondary beams connecting the trussed arches horizontally curve downwards like ribs to match the shape of the pneumatic ETFE cushions forming the roof covering itself. Aluminium sections fix the cushions to the top chords of the trusses, which at the same time form the joints between the separate cushions.

The timber space trusses supporting the roof would not have been technically and economically possible without using CNC machining centres. Neither the top and bottom chords, which are round members but at the same time have a linear curvature, nor the connecting members between the top chords could have been produced with standard woodworking plant. The small difference between arch members with a rectangular or round cross-section calls for a radical change in the production technology because the free-form surfaces that ensue due to the double curvature require large-format, multi-axis CNC machine tools for their fabrication.

Vitam'Parc Leisure Centre, Neydens, Haute-Savoie, France; architect: L 35 Arquitectos, Barcelona; structural engineers, timber structure: Charpente Concept France, Perly. 2007–2009.

for the concrete structures of the 20th century, which allowed structures like those of Pier Luigi Nervi to be built for the first time, so modern glued laminated timber can also be seen as a technological development stage that has a considerable influence on the aesthetics of its time.

The artistic opportunities presented by glued laminated timber in terms of the dimensions and architectural vocabulary of timber structures have been joined in recent years by another technological development that, in conjunction with glulam, has totally redefined the aesthetics of timber construction: computer numerical control (CNC), i.e., computer-aided manufacture.

The timber workpieces made from glued laminated timber are assembled to match the desired final shape approximately and then cut to the exact form in a second operation using CNC machine tools. The manufacturing process is therefore divided into a manual part, in which the basic shape is produced using traditional methods, and a highly automated, often very complex part in which the computer-controlled machine tool creates the geometric form required. The possibilities of CNC machine tools are manifold, ranging from three-axis CNC machining centres that can machine individual workpieces in the x, y and z directions right up to five-axis CNC plant that owing to the two additional degrees of freedom at the head of the tool can carry out highly complex three-dimensional machining operations with undercutting and hollowing-out. Saws and drills can also be mounted on the tool head, not just routing tools, which multiplies the machining options.

Cutting solid wood to size and shape is particularly simple and gentle on the tools. Therefore, this advanced form of shaping materials has become estab-

The total height of the tower is 66m. Each main column comprises six 20 × 20cm glued laminated timber members. Constructing the tower in glulam enabled it to be pre-assembled on site in four segments, which were then lifted into position by crane. "bahnorama" tower, for viewing the redevelopment of Vienna's main station, Austria; architects: RAHM-Architekten, 2008–2010.

“bahnorama” tower. View from north-west.

lished as an economic alternative for timber construction especially.

When it comes to very long spans, there is an economic limit to the beam depth/span ratio of glued laminated timber beams, even in the case of optimum design and construction. Besides structural assistance in the form of steel ties (trussed beams), it is also possible to employ concealed composite solutions, e.g., timber with synthetic, glass or carbon fibres, to improve the structural efficiency of glued laminated timber beams. To do this, carbon, glass or aramid fibres are glued in the tension zone of a glulam beam. The fibres are attached parallel with or perpendicular to the grain, either as a unidirectional bundle of fibres or as a flat mat embedded directly in the polyurethane adhesive between individual laminations. Consequently, the adhesive bonds the laminations together and at the same time functions as a matrix for the fibre material. Other advantages of this combined gluing are the good force transfer from the fibres to a large area of the timber and the fire protection afforded to the fibre-reinforced timber by the surrounding timber. Laying the synthetic fibres parallel with the grain can bring about considerable improvements in the tensile strength of the timber member and hence its flexural strength.

A study by the University of Natural Resources and Life Sciences in Vienna reports that the combination with synthetic fibres can improve the flexural strength of glued laminated timber by 70% and increase the flexural stiffness by 15%. Introducing synthetic fibres perpendicular to the grain almost doubles the ultimate load when compared with a conventional glulam element. In both these cases, the composite section subjected to an ultimate load exhibits a distinctly ductile behaviour, i.e., exceeding the permissible loads results in permanent deformation (yielding), but the composite section remains intact. There is no significant drop in the stresses in the cross-section, as is the case in the brittle failure of steel or, indeed, timber in tension. The amount of other material introduced is very low in all instances: at less than 0.7% by weight, the proportion of fibre-reinforced polymers is negligible.

Even though it might seem absurd at first glance, creating a composite material from two fibrous materials (wood and polymer) does prove to be a good choice. Here again, the principle of homogenisation applies: in practice, knots, sloping grain and the anisotropy of wood (i.e., behaviour dependent on direction) limit the excellent material properties that are theoretically possible. Adding synthetic fibres not only achieves the strength values guaranteed numerically, but in the case of strengthening perpendicular to the grain, can also achieve flexural stiffnesses and ultimate strengths that are otherwise impossible with glued laminated timber construction. We can expect more from timber/synthetic fibre composite technology in the future, primarily for long-span structures and especially when we consider that less than 1% of composite material can increase the ultimate and flexural strengths by as much as a factor of two.

Kerto-S laminated veneer lumber (LVL), paritcularly suitable for long-span beams.

Parallel strand luber (PSL).

Veneer and fibre materials

New developments in solid timber materials are joined by further developments in veneer and fibre materials. These are being used increasingly in long-span timber structures and likewise define a radical, new aesthetic in structures made of wood. An early form of this wood-based product was the plywoods that began to appear around 1850 with the advent of veneer production and

An interesting recent example of a long-span structure made from laminated veneer lumber (LVL) was the 2005 Serpentine Gallery Pavilion designed by the architects Álvaro Siza and Eduardo Souto de Moura. The structure for the pavilion was made from LVL assembled from two-bay elements in a way similar to the Zollinger form of construction. As with the Zollinger method, no bending moments were transmitted at the connections between the timber members, which resulted in a technically straightforward node design with very simple geometry. However, the dimensional stability of the individual elements was especially important for this lattice shell typology. Disparate shrinkage behaviour in the longitudinal and transverse directions, as is the case with solid timber, would have resulted in deformations, restraints, even damage to the roof covering. Plywood reacts to moisture changes consistently in all directions, and with its very low shrinkage (0.4% in all planar directions), together with the roof covering of polycarbonate sheets, it formed a construction that was very homogeneous in terms of its total deformation behaviour. Moreover, the high strength values of the LVL used allowed the connections to be designed as traditional woodworking mortise and tenon joints. The long-term durability of plywood (delamination of individual plies due to moisture, temperature stresses, etc.), which could be critical, was irrelevant for this temporary summer pavilion, especially as the polycarbonate sheets of the roof covering protected the timber against direct exposure to the weather.

Exhibition pavilion, Serpentine Gallery, London, UK; architects: Álvaro Siza, Eduardo Souto de Moura; structural engineer: Cecil Balmond (Arup), 2005.

the necessary rotary-cutting machines. Initially intended as finishes for furniture, veneers in the millimetre range of thickness soon began to be used to produce homogeneous plywoods with especially good loadbearing qualities. Veneers obtained from a tree trunk by rotary cutting and the subsequent cross-wise gluing under pressure of these millimetre-thick layers results in a material with properties that are closer to those of a modern polymer than those of the original wood. Laminated veneer lumber (LVL) is mainly used for sheathing and furniture. It consists of at least three plies of max. 7mm thick veneer glued together at 90° to each other. Boards are classed as suitable for use in dry, outdoor or damp areas, depending on their intended application and the adhesive used. Indigenous species such as beech, birch, poplar, spruce or pine are used for the veneers. Imported LVL products can also be made from Douglas fir or diverse tropical woods (e.g., limba, okoume, etc.).

The good dimensional stability of LVL compared with solid wood and its optional moisture resistance have encouraged the use of this wood-based product in long-span roof structures in recent times. However, the applications are limited to temporary structures or members protected from the weather.

Relatively new to the construction industry is parallel strand lumber (PSL), which is made from approximately 15mm wide by 3mm thick rotary-cut strips of softwood veneer (Douglas fir or southern yellow pine). The strips are glued together with phenolic resin parallel with the grain to form endless elements, which are cut to size accordingly. PSL is not produced in Europe; instead, it is imported exclusively from North America.

PSL is mainly used for straight members in loadbearing structures. The material exhibits excellent dimensional stability and strength and can therefore be used as an alternative to glued laminated timber. The standard beam size is 483 × 280mm; smaller cross-sections can be supplied by the manufacturer, larger ones can be assembled by gluing together several standard cross-sections. At 600–700kg/m^3, the density of PSL is very similar to that of the softwood used (Douglas fir) and is quite high when compared with fir (450kg/m^3) and pine (520kg/m^3).

Loadbearing beam made from parallel strand lumber (PSL). Housing estate in Waldkraiburg, Germany; architect: Hubert Riess, 1996.

Magnum Board®. Up to 10 plies of 25mm thick OSB/4 boards are clamped and glued together to create the storey-high wall panels up to 15m long and max. 250mm thick. In contrast to conventional OSB grades 1–3, the OSB/4 boards are characterised by their high moisture resistance and also by their suitability for heavy-duty loadbearing purposes.

Owing to PSL's high price (= approx. €1500/m³; for comparison: solid structural timber = €300–400/m³, glulam = approx. €1000/m³), it is used only occasionally even though its surface texture, colouring and homogeneity create interesting architectural highlights.

Modern wood-based products manufactured by cutting and chopping and reassembling with the help of binders date from the mid-20th century. Such methods enable the production of homogeneous, usually board-type wood-based materials. By changing the composition and using different manufacturing technologies, it is possible to generate many very diverse material properties, mostly new to the original wood, which then broadens the spectrum of potential applications. Oriented strand board (OSB), developed in the USA in the 1940s, is probably the best known product apart from the very common particleboard materials, which are primarily used in furniture and interior fitting-out. OSB unites the technological principles used for particleboard and plywood. The chopping of the raw material makes OSB similar to particleboard, although in contrast to the very fine chopping of the wood for the latter, OSB requires comparatively large chips (which are up to 100–200mm long, 50mm wide and several millimetres thick), which are then bonded together with a synthetic resin with the help of pressure and heat. The wood chips are arranged parallel, but in the outer zones of the board are arranged at 90° to this, which leads to an interlocking effect similar to that of laminated veneer lumber.

Although in the construction industry the customary OSB products used, at 12–30mm thick, are mainly used as roof and wall sheathing, a new development, marketed under the name Magnum Board®, makes OSB interesting for loadbearing components, too. Magnum Board® is an OSB product made from several plies of glued and clamped 25mm OSB 4, i.e., boards that can be used in damp areas and for loadbearing purposes. Between three and ten plies are glued together to produce elements 75–250mm thick. These thicknesses and the structural properties of the elements make them suitable for loadbearing wall or roof applications. We can assume that these OSB products will replace 3-ply CLT or plywood boards in one or other folded-plate-like

This fabrication building, completed in 1991, measures 30 × 60m and includes two overhead cranes, each with a safe working load of 6.3t, which adds up to a considerable load on the roof structure. The roof structure consists of parallel strand lumber (PSL) trusses every 4m arranged as a monitor roof, which allows plenty of daylight to reach the interior. PSL was also used for the columns. Owing to the high strength of PSL, a heavyweight glulam roof structure was avoided, which would have been more than 2.5m deep and therefore would have allowed less daylight into the building. Kaufmann Holz AG, fabrication building in Reuthe, Austria; architect: Hermann Kaufmann.

structures because their much more homogeneous material structure has considerable advantages over the other solid timber products. As of mid-2014, OSB/3, 22mm thick, costs about €10/m², which is much less expensive than the corresponding 3-ply CLT, which costs about €25/m², or plywood, which is about €20/m². However, when the boards are to remain exposed and therefore must be weather-resistant, the edges have to be treated because, like particleboard, the core of the board consists of coarse layers of chips which are very vulnerable to moisture. This treatment increases the price of these OSB products.

Another new development in material technology for timber construction is "wood modification", which means using mechanical, chemical or thermal techniques to change the structure of the wood in such a way that the result is a much better durability and dimensional stability than that of the non-modified wood. This is mostly achieved by applying heat to drive out the water molecules, which are in the cell walls of the wood, from the molecular structure, or to prevent their accumulation by chemical means. In another treatment method, the application of a waterproofing agent reduces the water absorption of the wood, which results in a better UV resistance. Even though this technology has not yet been widely used in timber construction, we can assume that the applications for long-span timber structures will increase as availability improves and the production costs fall.

Cross-laminated timber

Besides the well-known plywood products, other new solid timber materials have appeared over the last 20 years in particular which, like these, exploit the interlocking effect of cross-banded plies. As with glued laminated timber, strip- or board-type solid wood sections are glued together to form larger elements for construction. The best known of these is cross-laminated timber (CLT).

Cross-laminated timber, following on from glued laminated timber, represents a further step towards the homogenisation of the material properties of timber. The well-known undesirable properties of solid wood, such as splitting or distortion, are minimised in CLT. Several individual plies 10–35mm thick are glued together at 90° to each other to form panels between 42 and about 500mm thick. Lengths of up to 20m and widths of up to 5m are possible. An odd number of cross-banded glued plies is always used, so the grain in the outer plies is always parallel. Side boards, i.e., boards cut from the perimeter of the tree trunk, are mostly used, and primarily species such as spruce, pine, fir or larch. These segments generally exhibit better strength properties than the zones nearer the middle of the trunk, but the disadvantage is the greater deformations due to the effects of temperature and moisture, which makes them unsuitable for most purposes but is unimportant when the wood is used for CLT. The interlocking of the, mostly, three cross-banded plies of timber compensates for the different dimensioning depending on the direction of the timber.

Wood shrinks or swells very differently in the three principal directions (parallel with the grain, tangential to the annual rings, radially with respect to the trunk diameter). If the moisture content of the wood changes by 1%, the shrinkage of spruce in the direction of the grain, i.e., longitudinally, is lowest, at 0.01%, but is much higher in the radial direction (0.19%) and even more so in the tangential direction (0.36%).

Interlocking the timber plies by gluing them together at 90° to each other reduces the shrinkage and swelling to an insignificant amount, below the shrinkage figures of solid timber in the longitudinal direction (0.02% in the plane of the board, 0.24% perpendicular to the plane of the board).

The essentially more homogeneous properties of CLT compared with solid timber and even glued laminated timber represent a great advantage for long-span structures because the uniform shrinkage considerably reduces restraint stresses, splitting and gaping joints. On the other hand, it is necessary to take into account the considerable shear forces in the contact zone of the glue joint due to the interlocking effect between the plies. Owing to the interaction of moisture and temperature fluctuations, these joints can open up over time and destroy the bond between individual plies, which can lead to a multitude of different problems.

Furthermore, the interlocking effect does not mean that the CLT elements are isotropic (i.e., independent of

The element production building of timber fabricator Obermayr has a floor area of 3500m^2 and is an example of the use of wood-based products for large-scale, long-span structures. Both the pilaster-type columns erected on the inside of the walls and the two rows of internal columns, 27m apart, supporting the folded-plate roof structure and the crane rails, are made from conventional glued laminated timber. The external walls and the wide roof sections as well, whose alternating arrangement enables the inclusion of large, shed-type roof windows, are made from 44cm thick, thermally insulated timber sandwich elements with an inner sheathing of particleboard and an outer sheathing of OSB. Besides the low weight and the possibility of being able to include effective insulation in the core of the sandwich panel (in this case the 40cm cavity is filled with rock wool), it is the good structural efficiency of the building that is remarkable: on the east side the roof cantilevers 18m without intermediate support, and stability in the longitudinal direction is guaranteed by continuing the wide roof sections down to the ground at an angle. Diagonal steel bracing forming a structural connection between the upper and lower roof sections also contributes to the highly efficient design. With the diagonal steel ties, the result is a torsion-resistant structure, similar to a space truss, which permits a very large area of glazing and hence an excellent level of daylight in the interior.

Obermayr production building, Schwanenstadt, Austria; design: F2 Architekten, 2005.

Cross-laminated timber boards as roof elements to the G3 shopping centre in Gerasdorf, Austria. The great advantage of CLT over other wood-based products with a parallel grain structure is that splitting is rare and the shrinkage is equally low in both board directions; design: ATP, Architekten und Ingenieure.

direction) in all respects. Owing to the odd number of plies, there is a principal stress direction that corresponds to the direction of the grain in the outer plies. If elements are to span in one direction, then this principal stress direction must be considered during erection.

As there is usually a clearly defined principal stress direction in the case of long-span structures, the benefits of using CLT are limited to a few special cases in which there is either no clear principal stress direction or substantial shear forces or bending moments occur in the other direction as well. Folded-plate structures,

Another example of the use of CLT is the folded-plate structure for the small chapel belonging to the deaconess parish of St. Loup in Pompaples, Switzerland. This building is being used temporarily until the renovation work at the parish church is finished. Based on the principle of origami folds, the chapel was developed by the architectural practice Localarchitecture working together with the Laboratory for Timber Constructions (IBOIS) at École Polytechnique Fédérale in Lausanne (EPFL), and was built in 2008. For this small chapel with a floor area of 130m² and designed to hold 100 people, 40mm thick CLT is adequate for the vertical elements, 60mm for the horizontal ones. These are joined together with perforated plates and screws. Left untreated on the inside, the outer faces were finished with just one layer of waterproof sheeting plus a layer of 19mm impregnated three-ply core plywood. The two gable ends of the chapel are closed off with polycarbonate sheets fitted into irregular wooden frames. The opposing folds of the structure consisting of 92 different facets enhance the spatial effect of the chapel, lend the building the necessary stability and have a positive influence on the room acoustics. The sloping walls also assist with the roof drainage, which helped to keep the detailing of this building simple.

St. Loup Chapel in Pompaples, Switzerland; architects: Localarchitecture, Danilo Mondada and SHEL, in collaboration with the Laboratory for Timber Constructions (IBOIS) at École Polytechnique Fédérale in Lausanne (EPFL), 2008.

in which the planar elements are subjected to diverse loading conditions along the edges, represent another typical application for CLT.

There are many different products with composite cross-sections on the market. These are in the form of hollow, ribbed or box sections (e.g., Lignotrend®, Lignatur®, Steko®, etc.) whose individual elements in some cases exploit the interlocking effect due to cross-banded gluing. These products represent standardised solutions for certain applications and are characterised by a high degree of prefabrication and system integration, with correspondingly little work being needed on site.

Edge-fastened timber elements, with their upright strips glued, dowelled or nailed together, are essentially based on traditional models. They are mainly used for multi-storey buildings where acoustics, fire protection and thermal mass are important factors in the design.

Composite cross-sections that can be used for beams and columns include the four-piece section made from a quartered log which is squared but left with wane on one arris. The four sections are then glued together reversed so the heartwood is on the outside and the wane in the middle. This results in a diamond-shaped longitudinal hole in the middle of the section which reduces the weight of the section without diminishing its load-carrying capacity.

Summing up, it can be said that recent decades have seen a series of new developments in the field of wood-based products and that these have revolutionised the possibilities for timber construction. Besides the very familiar glued laminated timber products and their modern derivative, cross-laminated timber which can be used in highly complex structural systems (and whose main distinguishing feature when compared with glulam is its isotropic properties), it is primarily the fibre-bonded wood-based materials that need to be mentioned and will open up unforeseen opportunities in the future. These board-type materials with even more pronounced isotropic behaviour render possible very lightweight, highly efficient folded-plate structures, panel-type structures or rigid timber frames – all forms of construction that have already supplied a number of interesting examples. In the long term, these board-type wood-based products will change our view of timber construction from the stereotyped uniaxial design logic based on linear members to a biaxial design aesthetic based on planar members. We have witnessed this trend in the construction of multi-storey buildings for a number of years and it is now making the first inroads into engineered timber structures as well.

Glued laminated timber construction will also experience a second major renaissance. Combined with CNC production, which has made incredible technological advances in recent years, it is now possible to build architectural and artistic solutions that just a few years ago would have been possible only through extremely elaborate and costly manual operations. The homogenisation of the wood by way of prior grading, removing imperfections, finger-jointing and gluing in the production of glued laminated timber represents the foundation for further advances in high-quality CNC machining of timber and therefore supplies the essential raw material for the cutting and shaping operations of modern fabrication machinery. This new material aesthetic with its geometry not limited by planar or linear notions will remain the province of timber construction for the time being because hardly any other construction material is so kind to the tools and so easy to work as wood.

Five-storey passive-energy apartment blocks employing cross-laminated timber. "Zur Börse" client consortium project, Berlin-Prenzlauer Berg, Germany, during construction; architects: Müller Büro, 2010.

Lignotrend® floor panel.

CNC production for timber structures

Simone Jeska

The digitalisation of manufacturing processes and the associated continuous, digital process chain from design to production have not only brought about fundamental changes to manual operations, but are also having a lasting influence on architecture, structural designs and construction workflows. Much has been written about the epochal change from the industrial age to the information age. This change has also been felt, albeit with a slight delay, in fabrication for timber structures, which is experiencing a quantum leap comparable with how mechanised production replaced manual working in the 19th century.

In his doctoral thesis, Christoph Schindler describes how digitalisation has resulted in fundamental changes in production[1]. Making reference to Akos Paulinyi's writings on the history of technology, in which the tools of the various epochs are described and divided into hand tools and machine tools, Schindler introduces the expression "information–tool–technology", which he says marks a second interruption in this history[2]. Specific examples will be used here to illustrate how the various production techniques influence the design and hence the architecture. In timber-frame buildings where the loadbearing timber structure was worked manually with hand tools, every component is unique and every component is individually joined to the next component; accurate connections were produced by using one component as a template for the next. The introduction of steam-driven machines and the mass production of nails standardised the working processes, the components and their connections. Hand-made wood joints such as scarfs, laps or dovetails could not be produced by machines and so were replaced by connections with metal fasteners. The outcome of this type of production was the prefabricated balloon-frame buildings in the USA, which made use of standardised elements and connections that could be mass produced.

Shifting the information to the machine level combined with differentiated, flexible tool guidance enables more complex geometric relationships to be embraced and produced once again. This is because the machine not only processes the dimensional data for the components, but also links this information with details of each component's position in the overall structure, and places this in relation to the adjoining components in each situation. Changes to the information – the dimensions or the geometry of the components – no longer means extra work, and the upshot is that every component can be manufactured as a one-off, like when using hand tool methods, without incurring extra costs. This means, on the one hand, it is possible to handle free-form loadbearing structure geometries with a multitude of differently shaped members and, on the other,

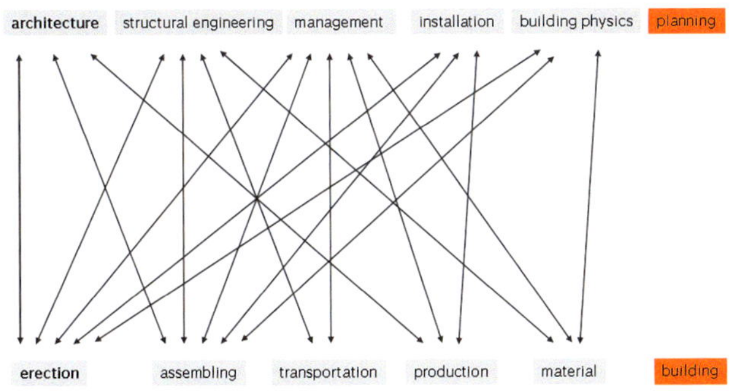

Considering digital fabrication as part of a digital process chain alters the interdependencies in the construction process.

1 He distinguishes between three production techniques by means of three factors: material, energy and information – material is processed using energy on the basis of information. The doctoral thesis was submitted to Prof. Hovestadt, ETH Zurich in 2009. See also: C. Schindler: *Die Mittel der Zeit – Herstellungsinnovation im Holzbau,* Arch+, No. 188, 2008.

2 Whereas with hand tools the material is shaped manually with muscle power, industrialisation saw machine tools take over the holding and guiding of the material and the tool, which therefore replaced muscle power, although people still controlled the machines. The introduction of the computer established information tools, which in addition to holding and guiding the material and the tool, also control the machine. The machine coordinates the processing of material and information.

that complicated connections, previously the province of manual working, are again becoming significant. In a similar way to the frequently heard "mass customisation"[3], the quasi-industrial production of individual timber structures is now possible.

The long-span single-storey timber shed structures popular with architects since the mid-1990s represent the manifestation of this new method of production. Sports halls with curving glued laminated timber beams and thermal baths or salt stores with spans of up to 120m are just a few examples of these new engineered timber structures which could not have been built without precise, efficient fabrication and effective connections.

Machining the components

The introduction of digitally controlled production plant in the mid-1980s has gradually changed timber construction[4]. At first, the new tools were used to increase the accuracy of fabrication and speed up operations; holes and cutouts could now be positioned very accurately, which had an effect on connection technology. The result of this development was connections with components joined together by steel dowels and steel plates let into the timber, which required an exact match between the pre-drilled holes in the timber and the punched holes in the steel plates[5]. The new connections had a direct influence on productivity and also the spans of the structures. Whereas at the start of the 1980s, the design and detailing of a single nailed joint in a truss took several hours[6], now only a few minutes were needed and parts lists could be produced as well.

Another leap in the development and production of timber structures was the arrival of five-axis CNC machining centres with automatic tool changers in the early 1990s[7]. The new tools could move in two directions simultaneously, which meant that curved geometries and angled or curved notches on sloping surfaces could be produced accurately and efficiently.

Owing to the constant development of the machines and their software, CNC woodworking machinery became the equivalent of the "intelligent" robots of other industries.

Fitted with swivelling and rotating units capable of movement in the linear X, Y and Z axes (or parallel with those axes) and in the rotary axes[8] A, B and C, the machine can plane, rout, drill, groove, saw, gouge, rebate and grind several workpieces at once if necessary, and there are practically no limits to the shape and orientation of the cuts, notches, tenons and holes or the workpiece itself. In doing so, it is not only the tools that move, but also the clamped workpiece. According to given priorities corresponding to the requirements of the machining process, the machine selects and changes the tools and machines the wood on all four sides.

In addition, the fabrication plant selects the workpiece to be machined from a stack of wood and positions it on the machine. Laser measurements of the exact position and dimensions of the initial workpiece

Six-axis robot with pivoting arm, University of Stuttgart, Germany.

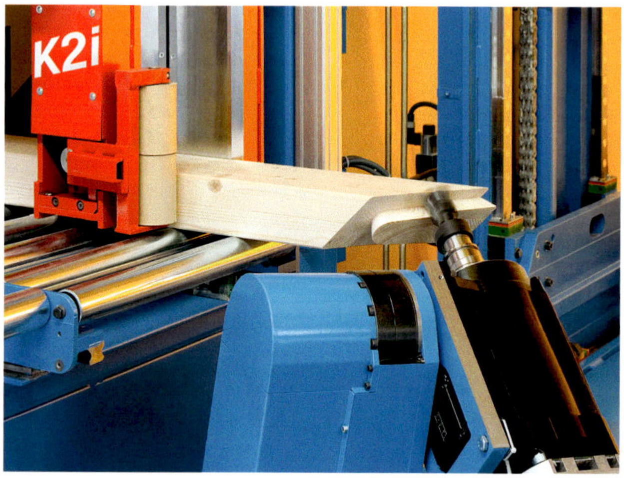

K2i woodworking machine with a six-axis robot unit, Hundegger company, 2011.

are sent to the computer so that the machining data can be adapted. At the end of the fabrication process, an integral numbering system (which reacts automatically and "intelligently" to changes) labels the workpiece so that its exact position within the total structure is defined (the erection of members or panels, all different, to form a structure is impossible without a clear system of identification) and allows the finished components to be stacked in sorted groups. Plant for manufacturing board-type workpieces in prefabricated construction also apply adhesives and screw, nail or staple the elements.

Each machine is controlled by an integral computer that imports the data from the CAM software. The CAM software converts the CAD drawings into production data[9], specifies the machining steps and uses nesting routines when cutting board materials to size.

Despite this automated high-tech production for timber components, the cutting and assembly of double-curvature components, which must be machined from all sides[10], is still a challenge. Just the clamping of the elements alone is difficult because when producing the curvature in the second plane, the workpiece no longer has a flat bearing surface. With dimensional tolerances of individual members in the region of tenths of a millimetre, the positioning of the workpieces and the precision of the tools are crucial. Currently, accurate positioning of the members in the plant makes use of CNC-fabricated templates, the making of which is already integrated into the design process. As an alternat-

The fish-belly glulam trusses have individual forms to suit the span (33–38m). Ice rink, St. Pölten, Austria; architects: Sam/Ott-Reinisch; timber construction: Glöckel Holzbau GmbH, 2007.

3 The term "mass customisation" – the industrial mass production of individual products – appeared in the late 1980s with the digitalisation of design and production processes.

4 The introduction of the HP 85 in 1980, the first computer with an integral monitor to be launched on the market, gave users the chance to display graphic elements. CAD programs and CNC machinery for woodworking were first seen in 1984 in the shape of the SPL 270-1 made by Burmek, a Swedish company; the German company Hundegger sold its first P8 machine in 1985.

5 Hermann Blumer recognised the opportunities for digital machining in timber construction in the early 1980s and developed the Blumer System truss. This new form of connection was based on further standardisation, so the design of the connection could be carried out by a specially developed program.

6 Up until the mid-1980s, it was still common in carpentry shops to produce assembly drawings and mark out the assembly on the floor.

7 The five-axis CNC Lignamatic was developed by Krüsi in Switzerland at the end of the 1980s; the prototype was used by Hermann Blumer in 1990. The main differences between that machine and today's Lignamatic are the computing capacity of the control (the prototype had a memory capacity of 800MB, whereas the memory of the new machine is measured in terabytes), the speed of the machining and the more efficient software.

8 Rotary axes A, B and C are axes of rotation about the principal axes X, Y and Z; using the rotary axes B and C, the tool can be moved at any angle (e.g., perpendicular to a curved surface).

9 The software uses the component geometries to define the geometrical and technical data, e.g., choice of tools, feed rate, rpm, orientation in space, movement directions, sequence of operations, types and positions of clamps.

10 The workpieces are always supplied oversize and machined on all sides in order to guarantee dimensional accuracy.

ive, the geometry and position of each member can be determined by the machine itself with the help of laser measurements and used for the various machining steps once the data has been converted. The positioning and fixing method chosen, the vibrations and deformations of the workpiece during machining and the use of curved workpieces, whose dimensional stability generally varies, have an influence on both the dimensional accuracy and the machining sequences and strategies and must be allowed for in the planning.

Building with straight members

Whereas changes due to the digitalisation of production were initially apparent mainly in the precision and rate of designing, calculating and fabricating, the effects are now evident in the three-dimensional complexity and/or individual geometries of the members and their connections. In particular, structures made up of large numbers of individual straight members can now be prefabricated economically from glued laminated timber elements. This development is readily apparent in the timber domes to salt stores, known as "Saldomes", in the form of gridshells in double curvature[11], the rediscovery of the Zollinger form of construction[12] and the construction of timber roller-coasters.

In the past some roller-coaster members had to be reworked on site and adapted to the circumstances. Today, however, the geometries of the notches in the members as well as the double-curvature rails[13] can now be determined accurately and prefabricated economically. The effects manifest themselves in the increasing size and spectacular appearance of the latest roller-coasters. The same is true for the Zollinger form of construction, which has been enjoying a comeback since the turn of the 21st century due to the economic and precise prefabrication of the glued laminated timber members. A total of 1794 curved glulam members (20 × 75cm) with nine different geometries had to be provided with sloping ends and trimmed top sides for the barrel-vault roof to an exhibition hall in Rostock, Germany. Designed with rigid nodes, these structures made from short straight members now provide roofs spanning up to 68m over trade fair halls and railway platforms.

Whereas the repetition rate of the rib geometries is relatively high for symmetrical barrel-vault roofs, the lattice shells of the free-form, double-curvature roofs over the thermal baths in Bad Orb and Bad Sulza[14] are

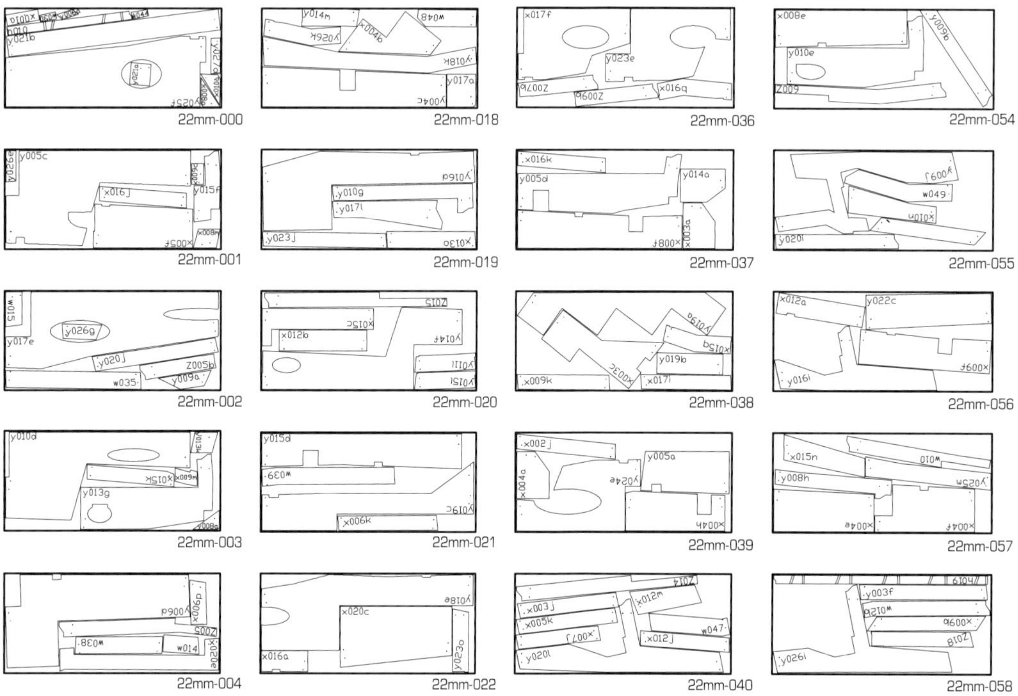

A program normally used in the textiles industry for cutting fabrics to size was modified to position the individual members on the standardised board material to achieve minimum wastage. "Burst", North Haven, Australia, System Architects, 2006.

made up exclusively of ribs that are all unique. Likewise based on the Zollinger form of construction, the approximately 1.80m square grid is made up of short, single-curvature members, which are about 3.60m long and therefore span two bays in each case. The curvature to the top face follows the course of the roof surface geometry. However, 12 different radii of curvature between 5 and 250m were defined for the underside, which means that the cross-section of each individual rib gradually tapers towards the end of the member (the camber in the middle of the member enables the ribs to be connected at the nodes; see the section on connections for further details). In addition, each member has rebates along the top edge for fitting the acoustic ceiling panels. This resulted in 682 complex and individual member geometries.

The recently completed projects with free-form, double-curvature structures or surfaces, such as those at the golf club in Yeoju, the Centre Pompidou in Metz or the theatre and concert hall in Kristiansand, represent both the culmination of digital production techniques and the prototypes for a continuous digital construction process chain on the basis of parametric models[15].

11 Up until now, double-curvature gridshells and lattice domes were either built by assembling layers of flexible boards or battens, which were bent to their permanently curved form on site with the help of templates or falsework and then fixed, or by combining straight bars arranged in a polygon.

12 Whereas in the past structures built using the Zollinger form of construction achieved spans of up to 40m, the roof structures to the trade fair halls in Rimini (2001), Rostock (2002) and Friedrichshafen (2002), and the roof over the railway platforms in Kassel have reached spans of up to 68m by using a method similar to Zollinger.

13 In 1999, structural engineer Werner Stengel registered a patent for producing the rails from prefabricated, glued timber sections which are mounted as complete rail segments. The prefabricated, double-curvature timber rails made from laminated veneer lumber were first used for the Colossos timber roller-coaster in Soltau in northern Germany.

14 Such a free-form lattice shell was built as long ago as 1999 to serve as the roof structure for the thermal baths in Bad Sulza. In those days, the ribs of the loadbearing structure were fabricated by Burgbacher on a three-axis CNC machining centre supplied by Hundegger (P10) and had to be finished by hand. The double-curvature edge beams were manufactured entirely by hand. The roof structures over the thermal baths cover areas of 2200m^2 (Bad Orb) and 1750m^2 (Bad Sulza); the shells are supported on double-curvature edge beams with eight and nine support points respectively and span up to 45m.

15 This production method had already been tried out successfully on smaller projects, e.g., the BMW Pavilion for the 1999 Frankfurt Motor Show, designed by Bernhard Franken and ABB Architekten, or Ludger Hovestadt's ESG Pavilion. The parametric model for the shape of the BMW Pavilion used a program from the film industry; the loadbearing structure consisted of a lattice-type aluminium rib structure on which the envelope of bent PMMA sheets was attached with discrete fixings.

Modern machines for working wooden board materials use nesting routines to optimise the use of the area and maximise the use of the material. The example is taken from the Hundegger company SPM-2 machine.

Today's complex free-form surfaces can no longer be processed using conventional design and production methods because they cannot be reproduced or calculated exactly by means of two-dimensional drawings (plans, elevations, sections). Furthermore, the typically large numbers of individual components that have to be coordinated cannot be produced within a reasonable time-frame, nor for a reasonable budget. A three-dimensional representation of the geometry, the direct transfer of the data from design to production and the links to efficient machines are absolutely vital for the realisation of such projects.

A parametric model[16] is developed as the basis for the design work, calculations, production and erection of such projects in order to be able to represent and, in the end, fabricate the constantly curving geometries. The advantage of the parametric drawing, which has been used as a design and planning tool in mechanical engineering and industrial design for many years, is that the form is not defined by means of absolute values, but rather by means of certain rules that determine the configuration of the form and the relationships between individual parts, depending on defined parameters. This means that the effects on the geometry of the individual components due to changes to the overall geometry or details can be ascertained, calculated and taken account of – all automatically. The entire system reacts to suit a single change at a single point. Planning steps, e.g., specifying the geometry of components and nodes, determining their position in space and their relationships with one another and the building grid-lines, are automated.

Despite automation, however, every step in the design and calculation work for these non-standardised geometries presents a challenge and calls for custom solutions. For example, the structural calculations to determine the internal forces and dimensions for the 45 000 or so members of the timber structure in Metz included 96 wind and 80 snow loading cases, which resulted in 1.2 million potential combinations having to be considered! In order to be able to carry out such huge numbers of calculations in a reasonable time, new, multi-stage programs had to be developed and used. As the connections and nodes as well as the segmentation of the members influence the stiffness of the loadbearing structure, every part of the structure had to be drawn separately – every chord, every shear cleat and every node was designed with its corresponding orientation in space – in order to specify the final dimensions of the components. Elaborate optimisations and calculations were performed in order to define the segmentation of the members.

This approach helped to reduce the cost of materials, minimise the number of splices, standardise the fabrication processes and speed up production. To determine the optimum geometric relationship between initial workpiece and finished member, or member segment, the clamping volume and the associated machine hours were calculated as well as the costs for the splices compared with the cost of producing curved workpieces – shorter segments lead to costs in the production of connections, longer segments must be fabricated from more expensive, curved workpieces. In all this, it was also necessary to allow for the maximum

16 In a parametric model, the form is not defined by way of specific dimensions, but rather by way of specific rules which determine how the form is configured depending on certain parameters and how the individual parts behave with respect to each other. Changes at a single point cause the entire system to respond to suit, and the steps in the design are therefore automated.

The gigantic frameworks for timber roller-coasters can involve the assembly of up to 120 000 straight timber members. Owing to the overlaps and intersections of the structure, those members require various skewed or diagonal cutouts and notches, which means that most of the straight members also exhibit individual geometries. “Mammut”, Tripsdrill Theme Park, Germany; Ing.-Holzbau Cordes, 2008.

The double-curvature lattice shell is made up of short glulam members, each one of which has a unique geometry. Toskana thermal baths, Bad Orb, Germany; architects: Ollertz Architekten; structural engineers: Trabert + Partner, 2010.

fibre cutting angle of 5°, required by the structural design, which was calculated at 25cm intervals along every member. During the optimisation process, a permanent retrospective link to the structural analysis was necessary and only after finally specifying the segmentation and completing the calculations was it possible to begin machining the workpieces.

Nodes

In a similar way to the rediscovery and further development of traditional loadbearing structures, traditional wood joints such as dovetails, scarfs, tenons and laps are experiencing a renaissance due to the digitalisation and automation of production. Not only do the new CNC-fabricated "old" wood joints improve aesthetics by providing alternatives to joist hangers, punched metal plate fasteners and steel plates let into the timber when building new roof structures and ensure an economic alternative to manual woodworking techniques when refurbishing existing timber-frame buildings, but they are also becoming established in engineered timber structures for the first time. The return to traditional wood joints is also reflected in research projects and undergraduate theses. In order to sound out the possibilities for a digital production chain from design to

17 ESG stands for "Endless Space Generated by individual sections". Totally in keeping with the mass customisation concept, a program was developed which provides users with a design tool for building a pavilion to suit the user's specific requirements. Internal (2 × 2.25m) and external (3 × 2.85m) rectangles create a frame within which the design variations can be developed.

18 Oliver Fritz from the Department of Architecture and Digital Media is implementing this concept, "Understanding Through Experimentation", together with Johannes Käferstein and Urs Meister from the Department of Structural Design.

This curving, double-curvature timber structure covers an area of 8000m^2. Centre Pompidou in Metz, France; architects: Shigeru Ban/Jean de Gastines; timber construction: Holzbau Amann GmbH; structural engineers: Ove Arup, Hermann Blumer, SJB Kemper Fitze AG, 2010.

finished product, Ludger Hovestadt and his ETH Zurich students developed the ESG Pavilion[17] in 2004 – a furniture-type pavilion made from wood and polymers. The timber structure consists of plywood assembled to form frames with a variable cross-section, which are then placed every 60cm and interconnected with plywood strips. The frame segments are joined by CNC-routed, glued dovetail joints to form a stiff frame; tenon joints connect the plywood strips to each frame. As a continuation of this experiment, students from the University of Liechtenstein[18] developed designs on the basis of parameterised, traditional wood joints according to the bottom-up principle: "...nodes were designed first, then frames and finally spaces."

Besides these intricate three-dimensional geometries for the connections, CNC technology also simplifies any form of individual node design. Instead of the customary steel plates, the connections between the members of the spatial, irregular truss for car manufacturer Wiesmann use 40mm thick plywood pieces, which are let into slits cut in the ends of the members and nailed in place. Owing to the irregularity of the structure, every piece of plywood is cut differently and every node has a different arrangement of nails. The individual nailing of the connections was made possible through the use of CNC-fabricated templates.

Free-form structures assembled from straight members result in complex, three-dimensional nodes

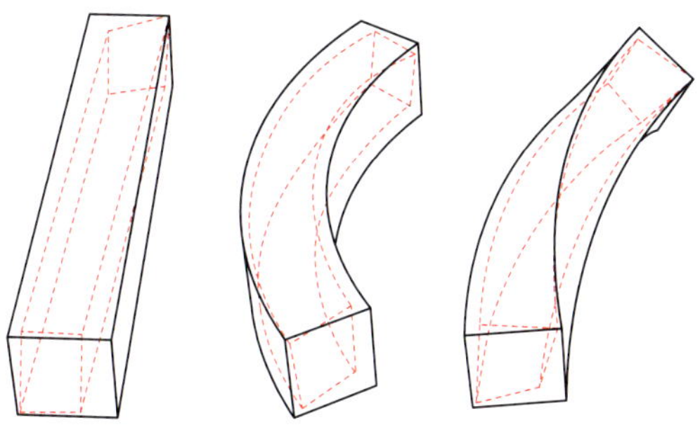

Segments are allocated to groups and assigned to different workpiece geometries while taking into account the radii of curvature and the raw materials available, the transport options, the size of the working area of the fabrication plant and the maximum length of the member segments.

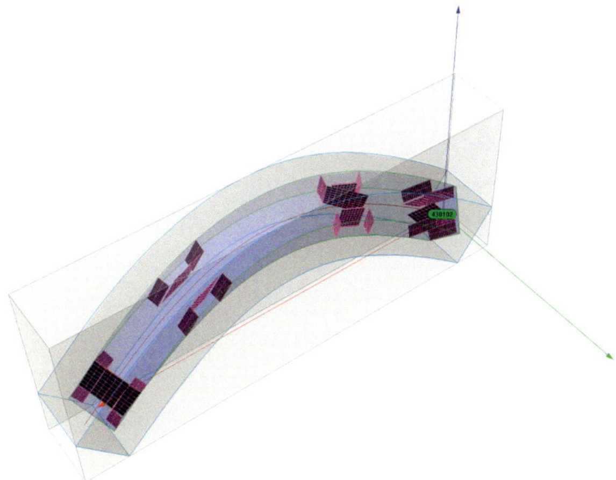

Digital representation of a member segment with notches. Golf club in Yeoju, South Korea.

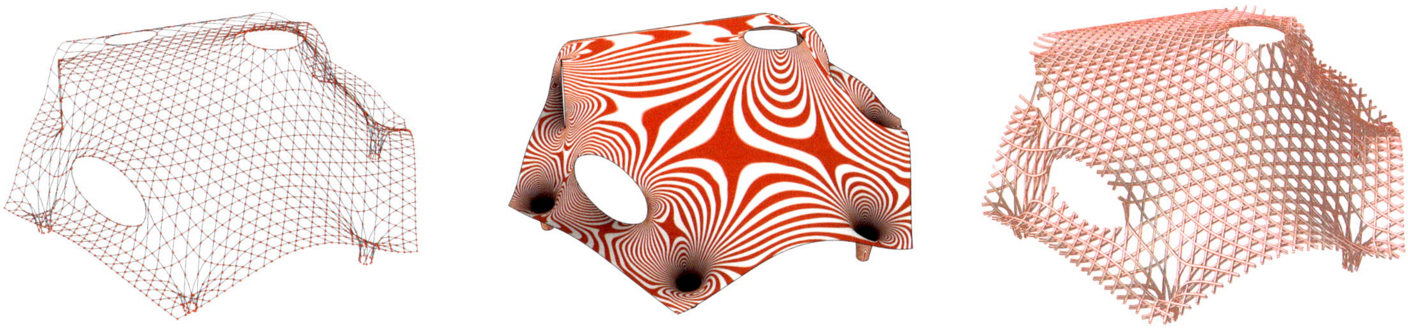

A continuously curving reference surface was developed from the 3D model with data regarding the positions of the nodes of the structure. Using the calculated member dimensions, the program designs the geometry of the double-curvature members along the single-curvature member axes and automatically ascertains the interface geometry of the nodes; geometry preparation: iCapp. Fabrication optimisation: designtoproduction.

requiring considerable input in fabrication and erection. Up until now, the connections for such straight members were either standardised and the members joined by way of geometrically complex steel parts (gusset plates, etc.) or the members were bent to suit on site before being fixed[19]. Parameterisation of the CAD data coupled with five-axis CNC machining centres enables the fabrication of such complicated three-dimensional connections; member ends are machined individually for the nodes and connected with standardised built-in steel fittings. This simplifies erection, reduces the tolerances of the finished structure to a minimum – the structure is almost like a piece of furniture – and encourages the development of new nodes and connections.

The complex loadbearing structure for the Centre Pompidou in Metz, France, resolved into six loadbearing chords, was achieved with simple connections using prestressed threaded rods (M24)[20].

Another example of this development is supplied by the nodes and connections of the roof structure to the clubhouse in Yeoju, South Korea, where traditional wood joints were used in an engineered timber structure for the first time. As in Metz, the free-form roof is assembled from a lattice of three beam axes, which are arranged in such a way that pairs of axes intersect; however, whereas the Centre Pompidou beams are resolved into groups intersecting at different levels, the beams in Yeoju meet in the same plane. The fire resistance specification prohibited the use of steel connectors at the nodes and so double halving joints were used.

To increase the bending stiffness of the structure[21] and simplify erection, the beams are in the form of pairs of members fitted together back to back with alternating notches. This paired arrangement of the members enables the connections to reach a degree of efficiency of 90% for bending actions.

The challenge here was not just verifying the load-carrying capacity of the connections[22], but also defining and producing the geometry of the intersections. This is because intersections between curved and at the same time twisted members result in HP surfaces – double-curvature hyperbolic paraboloids – instead of flat surfaces, and that means varying geometries, too, corresponding to the curvature of the members themselves. As the standard design tools for such structures were unsuitable, a program with universal rules for all nodes had to be developed so that the halving joints could be designed automatically in the 3D model[23].

At the end of these planning and calculation processes, which define the exact geometry of every single member, the component geometry of the members (there were 45 000 of them in Metz) is converted to data that controls the fabrication machinery. The detailed structural models allow any problems or ques-

The members to the roof structure for the golf club in Yeoju intersect in the same plane. The double-curvature beam segments have sloping, curved notches for the halving joints.

tions regarding details to be checked immediately during the course of the work, so that the necessary adjustments to suit the new circumstances can be carried out during production.

Building with boards

As with loadbearing structures built from straight timber members, structures made from planar, board-type elements can benefit from innovations due to the digitalisation of design and production. The result is customised wall panels and roof coverings or unusual lattice shells and grillages made from timber boards. Modular buildings, which originally relied on standardised components in order to achieve cost-effective production, can now be custom-fabricated just-in-time without causing additional costs – and there are no limits to the potential plan forms or internal layouts. Furthermore, any openings can be cut in the boards as required and the surfaces routed to match architects' designs, which is leading to a renaissance in decorative features.

"Burst", a beach house built in North Haven, Australia, in 2006, is an excellent example of the new generation of modular buildings. Skewed and tapering wall and roof surfaces made from coated, thermally insulated timber boards, which cover a large area of the building, leave us in no doubt as to the individual

19 Double-curvature lattice domes were mostly built using layers of material or grids of flexible battens (Multihalle in Mannheim, kindergarten in Triessen, exhibition pavilion in Nara, etc.).

20 In a series of tests in 2007 at the Higher Technical School of Wood in Biel, Switzerland, several full-size models with dimensions of up to 10 × 12m were built in advance in order to test the load-carrying capacity and stiffness of the connections and evaluate their structural efficiency.

21 Compared with a traditional halving joint, having two intersecting beams each with half the notch increases the bending resistance by a factor of three in a design with resolved beam cross-sections. At the same time, resolving the beams strengthened the respective principal beam direction.

22 Connected with the establishment of new, non-standard nodes is the determination of structurally relevant parameters by means of tests which then specify the dimensions of the built-in fittings or, as with the structure in Yeoju, the gluing. Extensive series of tests were carried out at Bern University of Applied Sciences for the structural connections in Metz and Yeoju.

23 CAD programs that present the connection details in parametric form have been available for timber construction for a long time, but they are unsuitable for working with free-form geometries. Therefore, all the design work for Metz and Yeoju was carried out with software from outside the construction industry, and that data then had to be transferred to the customary CAD software via new programs. In the case of Yeoju, designtoproduction supplied finished 3D models of all components (and for Metz the components were designed by the timber contractor with the help of plug-ins).

The loadbearing structure for the Centre Pompidou in Metz, France, has four curved and twisted glued laminated timber chords (140 × 440mm) intersecting on different levels at the nodes. Hexagonal timber dowels made from beech laminated veneer lumber ensured the dimensionally accurate yet straightforward erection of the geometrically demanding nodes and served as guides for the prestressed threaded rods that fix the nodes. The exact positioning and perpendicular orientation of the 15 000 dowels in the twisted and skewed surfaces was made possible only through the use of digital precision tools. Short screws and conical double nuts were used to fix the position of the layered chords.

designs possible with this modular system – an impression reinforced by the decorative floral motifs.

The principle of the production of a series of individual parts continues in the loadbearing structure, which consists of 25mm thick plywood ribs with different lengths and depths. These are connected using simple, X-shaped steel plates to form an irregular, diamond-shaped grid. The horizontal grid for the floor continues in the vertical and horizontal structures of wall and roof. Projections and returns have been created so that on the inside, the loadbearing ribs also function as frames for stairs, platforms, benches and built-in cupboards; outside, they define the form of the skewed wall and roof surfaces. In the end, 1100 plywood ribs – all with a custom geometry – had to be fabricated. A computer program used in the textiles industry for producing cutting patterns was used for the design and cutting to size. Based on a multitude of sections through the building along every grid-line, the software generated the geometry of the individual ribs, calculated the most economic use of the board material and sent the data to the CNC router, which fabricated, labelled and packaged the components in sets[24].

The little beach house not only signalled the emergence of a new generation of modular buildings, but also a new design principle. The design team exploited the advantages of the board material, which can be readily cut to any shape, in order to create an unusual, grid-type loadbearing structure providing more than just a supporting framework. This design and production method was taken up and transported to a larger scale for the Metropol Parasol scheme in Seville. The free-form lattice shell of Kerto-Q laminated veneer lumber (LVL) panels up to 28m above the ground is supported on six tower-like columns and covers an area of about 10 000m^2. The LVL panels are arranged on an orthogonal, 1.50m square grid and create a three-dimensional volume with free-form, undulating concave and convex surfaces. That meant 3400 different components for the ribs; and not only is the geometry of each one unique, but each one has an irregular form – there are no right-angles or straight lines. With a maximum depth of 3m and thicknesses ranging from 68 to 311mm due to structural requirements, the lengths vary between 1.50 and 16.50m depending on the curvature and the slope of the fibres (max. 7°). Rigidly connected with built-in steel parts, the timber lattice has spans of up to 40m.

The effect that the digital fabrication of series of individual members has on the architecture is not only apparent in the form of new, unusual structural designs. It is also evident in the configuration of façade and wall structures assembled from a multitude of small elements. One example of this trend is the double-curvature, wavy wall surface to the Kilden theatre and concert hall in Kristiansand, Norway.

The wall is made from about 14 000 tapering, twisted, oak boards that are screwed to a concealed framework of curved glued laminated timber beams. The consequence of the complex, irregular geometry of the surface is a huge number of separate pieces and individual connections – the twisted and at the same time tapering boards of the double-curvature wall result in a different connection geometry at every connection with the framework behind – which had to be described and fabricated, with all the details, in three dimensions. Precise labelling of the boards denoted their exact position within the total construction. Every board can only be used at a certain point in the façade, so a numbering system was crucial.

Digital fabrication as part of a continuous digital process chain and the trend towards assembling large timber structures, façades and walls from huge numbers of individually fabricated components are mutually dependent. On the one hand, parameterising the geometry simplifies the design work, but, on the other, it results in a change in the sequence of design and construction processes, whose interdependencies, conditions and interactions have to be redefined. Design, calculation, fabrication and erection, originally regarded

24 It was therefore possible to build the beach house (100m^2 floor area) for €150 000. A second prototype was built in New York in 2008 for the exhibition "Home Delivery: Fabricating the Modern Dwelling" – an event that demonstrated new visions for sustainable, prefabricated houses.

Decoration routed directly into the loadbearing structure represents a reinterpretation of traditional wood carving. New Monte Rosa mountain hostel, Switzerland, Gramazio & Kohler, 2009.

as separate processes in a chronological building process, culminate in the planning phase, which shifts the complexity and the majority of the work in that building process from the building site to the design office; a feature of this re-allocation is the comprehensive and complex data network during the planning phase. The complete structure already exists as a virtual model right down to the details before it is built and, as a physical model, only has to be assembled like a puzzle. Knowledge about what can and cannot be done in terms of erection, materials and fabrication have to be taken into account during design. The nodes and interfaces of the loadbearing structure and the erection pro-

Viewed from the road, the beach house looks like an irregularly folded block on stilts. "Burst", North Haven, Australia, System Architects, 2006.

cedures have to be known in advance of the structural analysis.

In addition, connections have to be as straightforward as possible, components must be unambiguously labelled to ensure their correct allocation and inspection mechanisms have to be in place to ensure exact positioning. When erecting free-form structure geometries in particular, establishing a reference system for setting out the position points on site is time-consuming, but this work can be minimised through maximum prefabrication. During pre-assembly, control lines have to be defined or templates set up in order to check the dimensional accuracy and the exact orientation of the members. Although the use of templates represents a reliable method, the complexity of the templates means it is an involved, time-consuming and costly method that is worthwhile only when the templates can be used more than once, as was the case for the roof to the golf clubhouse in Yeoju (see p. 109). Joining the components together, on the other hand, is comparatively unproblematic: general arrangement drawings with the members marked according to the numbering system define their positions within the total structure.

The projects described above, which were pioneers in exploiting the potential of CNC technology, at

Despite being part of a system, every façade panel is cut differently.

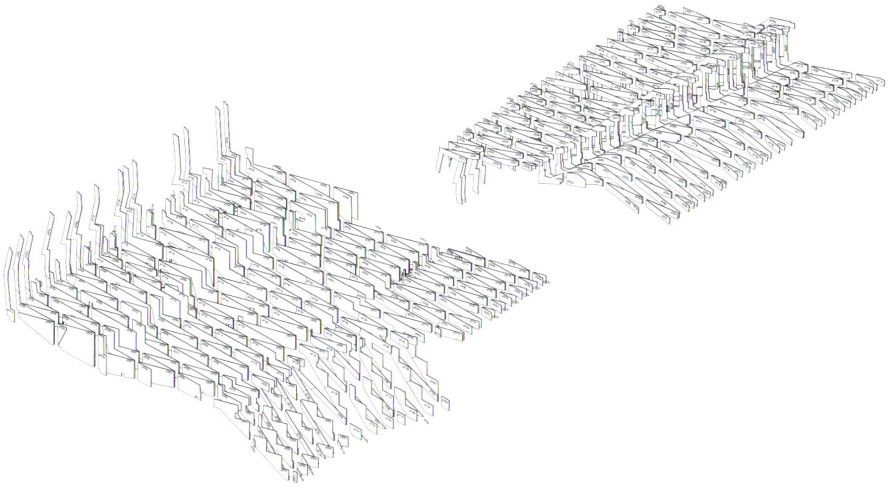

Owing to the different shapes of the boards, the loadbearing ribs at the same time form the supporting structure for the external stairs.

The production of L-shaped timber ribs for the floor/wall and wall/roof transitions automatically resulted in rigid corners. The cut plywood boards were easily assembled on site by students following structural location drawings.

the same time revealed the weaknesses and problems that can occur in the production process. The lack of suitable planning tools and interfaces led to costly custom solutions; programs from the textiles or automotive industries[25] had to be rewritten for these purposes, and interfaces[26] had first to be developed[27].

Despite these difficulties, the advantages of digital fabrication are obvious: In preparing for production, the provision of component drawings with the appropriate notches and holes for nodes and connections is automated and changes can be carried out at any time and allowed for automatically. Parts and materials lists for procuring the raw materials can also be generated automatically; the precise compilation and calculation of timetables for the individual operations allow the fabrication times to be determined exactly. Grouping together the necessary fabrication operations in one machine and the digital control of the tool positions obviate the need to measure, mark out, build gauges and move the workpiece from one machine to the next, which thus shortens production times enormously. In addition, the precision in fabrication, the production of one-off components at no extra cost and the opportunity of producing geometrically complex components are leading to innovations in timber construction.

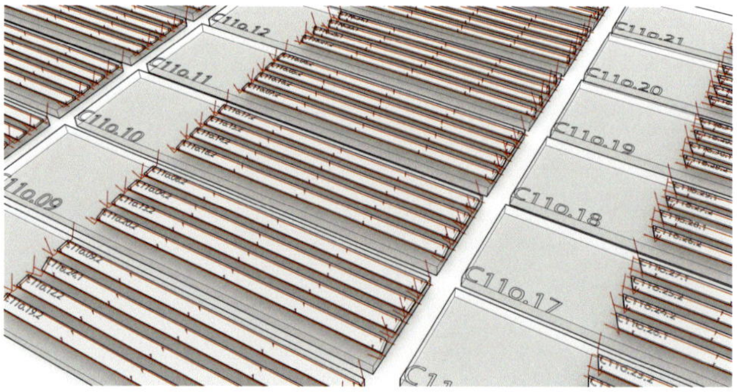

To enable the boards to be cut to size, the twist of every single component had to be reduced to a flat plane. The grooved glued laminated timber beams and the CNC-fabricated screw holes defined the exact position of every board.

The geometry of the wall surface changes from a straight eaves line 22 m above the ground to an undulating outline at the base a considerable distance behind the façade itself. Theatre and concert hall in Kristiansand, Norway, ALA Architects, designtoproduction, 2012.

25 Converting the three-dimensional DXF files of the design into a flattened, parametric NURBS model for the roof structure in Metz required the help of specialists from the automotive industry because the customary CAD programs had reached their limits.

26 `During the fabrication of the roof structure for the clubhouse in Yeoju, missing programming interfaces in the CAM software led to the curved components having to be programmed as separate pieces on the basis of the detailed 3D models in the CAM system.

27 The problems in CNC fabrication due to the lack of standardisation in the software interfaces between the CAD, CAE and CAM data, which lead to time-consuming, costly planning, were already evident in the early 1990s: the drawings prepared by the architect and structural engineer could not be imported into the system and so the working drawings and production schedules had to be prepared anew and the fabrication data recompiled. A research project on data transfer in timber construction, initiated by the German Association for Timber Research (DGfH), investigated this topic in 1995.

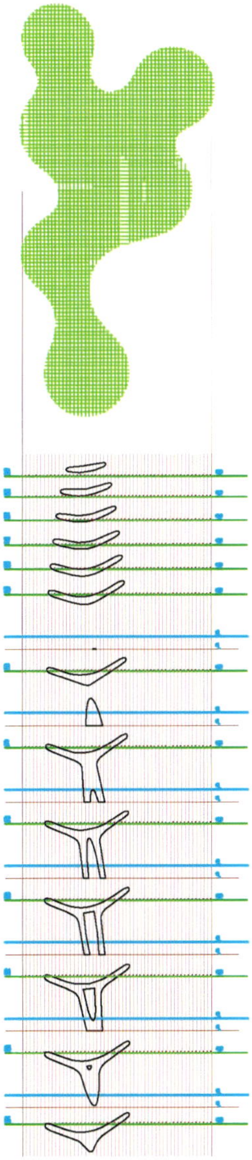

Free-form geometries are the result for the cutting of the individual panels. The panel geometries were calculated from the digital solid model according to the "egg slicer principle".

The gridshell of LVL panels is braced by steel diagonals. Metropol Parasol in Seville, Spain, Jürgen Mayer H., Arup, Finnforest Merk, 2011.

Experimental and temporary structures

Simone Jeska

Before new materials, new loadbearing structures, new connections and new production methods can manifest themselves in the built environment and establish themselves as recognised construction materials and techniques, they are often tried out in experiments or used for temporary structures. In this respect, wood, as an inexpensive and easily worked material, is especially suitable for experimenting with new ideas. Stacked squared sections and timber boxes create spaces that are positioned between art and architecture; they exploit the atmospheric effect and the cultural significance of the material and impress us through their simplicity.

The Swiss Pavilion at EXPO 2000 in Hannover employed stacks of squared timber sections to form a labyrinthine sounding box, whose aim was to seduce the senses. The pile of wood, which we associate with logs stacked in front of a country home ready for the fireplace, underwent a reinterpretation and led to irritation. An 8.68m high stack of two parallel lines of 100 × 200mm sections with transverse 45 × 45mm battens formed each semi-permeable wall of an internal

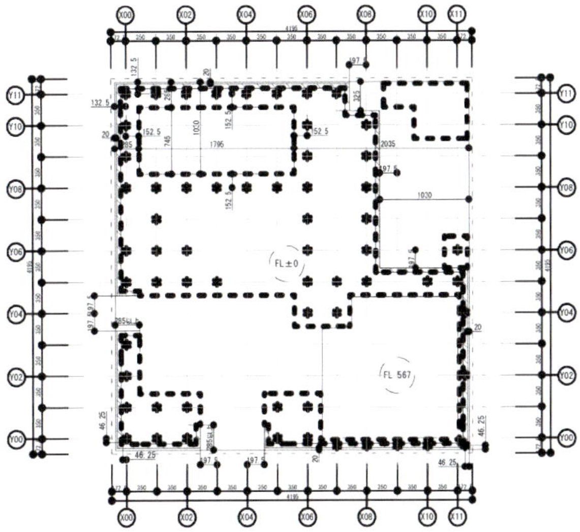

The pavilion at EXPO 2000 consisted of 12 structurally independent stacks of at least four parallel walls of stacked timber sections. Tension cables, with small cranks at their fixings, counteracted the approx. 60mm horizontal displacement of the stacks of timber sections – the result of shrinkage. Changes in height amounting to max. 23cm due to shrinkage and creep were compensated for by a tension spring at the top end of each cable, Atelier Peter Zumthor & Partner, 2000.

Holiday home in Kumamura, Japan. The small bathroom directly adjacent to the entrance is the only enclosed space in the otherwise open, flowing interior landscape. Services such as electric cables and heating and water pipes are concealed so that nothing disturbs the impression of a random pile of timber sections, Sou Fujimoto Architects, 2008.

layout characterised by aisles and courtyards. An interior without boundaries and accessible from all sides. Only the cafeteria and ancillary rooms were roofed over. Cross-beams linked the stacks of timber at a height of 6.25m. The flexural strength of those loosely stacked sections and the corners of the frame was achieved through vertical, prestressed steel cables anchoring the stacks to the ground every 3m.

The pavilion defied our standard notion of enclosed space; instead, we were given permeable layers of space that separated visitors more and more from the "outside world" the further they penetrated into the interior of the pavilion. As when entering a forest, the acoustic and visual references gradually disappeared until, reaching the heart of the edifice, an independent, atmospherically dense acoustic and visual space emerged: light from above infiltrated the roof-less pavilion, which became a sounding box for musical and vocal performances. In its primitiveness, the wood provided a bond with memories and at the same time became a multifaceted, ambiguous system.

Sou Fujimoto also exploits the primitiveness and purity of the material for his reinterpretation of a traditional log cabin. His cube-shaped holiday home (side length = 4.20m) in Kumamura, Japan, is formed by 191 cedar sections measuring 350×350mm, which are placed in 11 horizontal and 12 vertical layers. Each horizontal layer is different on plan and every vertical layer is different in elevation. This leads to a complex, cave-like interior space in which the stacked timber sections can also function as stairs, benches, tables, shelves or platforms. The apparently random stacking of the timber sections results in projections and returns, cavities and recesses that are practically climbed by the inhabitants in order to search for a suitable place for sleeping, eating, reading, writing or relaxing. This holiday home therefore defies our conventional understanding of architecture. With the exception of the small bathroom next to the entrance, there are neither separate rooms nor any areas or components with defined uses: the stairs can be used as table or bench, the wall as stairs or shelves. It is only the inhabitants who attach temporary definitions to the places. The squared section becomes a universal component, which is at the same time enclosure, loadbearing structure, internal structure and furniture. Therefore, the entire spectrum of the material with its physical and structural properties and surface qualities is exploited. Threaded steel rods (d = 18mm), which connect the sections vertically and horizontally, frame-less, single-glazed windows (6mm) and a frame-less pane of glass for the roof are hardly noticeable and reinforce the impression of a loosely stacked pile of wood.

Whereas in the case of the EXPO pavilion and the holiday home in Japan, the wood was/is used (almost) in its original form, the building material for the pallet

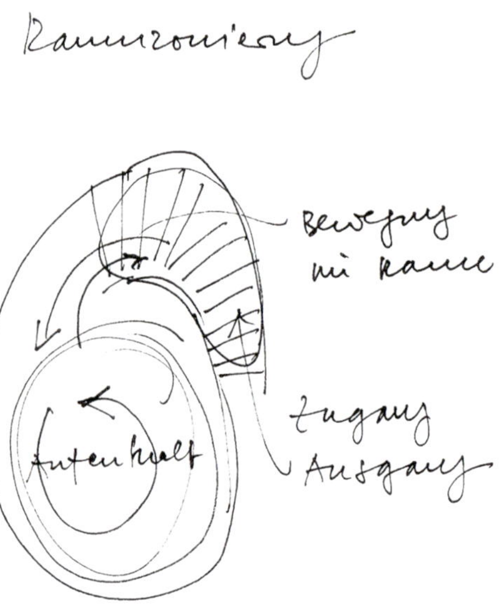

The amorphous pavilion for the 2005 World Ski Championships in Oberstdorf, Germany, was made from 1300 Euro pallets; Prof. Matthias Loebermann, Biberach University of Applied Sciences.

pavilion in Oberstdorf, Germany, employed industrially manufactured everyday objects. Euro pallets are unheeded "cheap products" in everyday use; but used as wall elements, their load-carrying capacity is confirmed and they undergo an aesthetic reinterpretation and upgrade. A sustainable design plus, straightforward assembly and dismantling by unskilled labour were the requirements for this pavilion, which was needed as a meeting point for athletes and the media during the 2005 World Ski Championships. Based on an amorphous plan form measuring about 8 × 18m, the pallets were stacked 6m high to form an organically curving enclosure with convex walls. The spiral concept behind the plan layout of the pallets of the lower layers produced an entrance that, like a snail shell, created a fluid transition between exterior and interior, and despite the lack of a door shielded the meeting zone from the weather.

The design itself was amazingly simple: Each layer of Euro pallets (1300 in total) was offset horizontally by half the distance between two runners and stacked set back or overhanging by a few centimetres. Ratchet straps, as used on heavy goods vehicles to tie down loads, were threaded through the pallets every 2.50m to create what was essentially a "prestressed" shell. The straps were tied to anchors in the ground. Despite, or perhaps precisely because of, this amazing simplicity, the mundane was transformed into the sublime. The wooden pallet as a building module was loadbearing structure, diaphanous envelope and decoration all at the same time.

Besides these experiments with enclosing spaces inspired by timber, the combination of new, digital design, calculation and production methods in recent years has led to a number of remarkable timber structures, especially at universities.

The result of the annual student competition for a temporary pavilion at the Architectural Association School of Architecture in London[1] was an unusual gridshell made from laminated veneer lumber (LVL) boards.

1 The pavilion was built in the summer of 2008 in Bedford Square, London. The design by students of the Architectural Association School of Architecture was supervised by Charles Walker and Martin Self, who were in charge of the summer pavilion program within the scope of the "Design & Make Studio" from 2005 to 2009.

Digital design methods were used to develop the structure of this gridshell made from about 650 vertical LVL boards with a light-coloured stain finish. The offset transverse ribs were screwed to the curving longitudinal ribs. Temporary pavilion, London, UK, students of the Architectural Association School, 2008.

The idea behind the design was to sound out how new technologies could be used in architecture; accordingly, the pavilion geometry was developed with the help of a 3D computer program and the structure fabricated with a CNC router. The pavilion was christened "Swoosh" by the students because on plan it was formed by two loops, reflected symmetrically about a point, which were inspired by the Nike logo. Above the double-loop plan, the lattice structure evolved through constant changes in height – from low-level benches at the opposite ends of the loops to cantilevering canopies that, in the middle, merge to form an arched walkway. The gridshell was formed by mostly multi-part curving ribs made from 51mm thick LVL boards, which radiate out from closely spaced steel fins at their base and are joined together by rectangular transverse ribs made from 27mm thick LVL boards. This unusual use of timber boards as the building material for a lattice shell structure, which would normally be built from linear-type members, had a famous role model in the shape of Álvaro Siza's Serpentine Pavilion. The playful, liberal treatment of the material and the design resulted in a curving, space-forming sculpture that demonstrated new artistic opportunities for timber construction.

The material behaviour of birch plywood, which is characterised by its flexibility but at the same time high strength, was the reason for devising a new type of structure for a temporary research pavilion at the University of Stuttgart[2]. A planning tool developed by the team itself enabled the design of a structure that exploited the elasticity of the material in order to create a stable, stiff system. In this bending-active structure[3], flexural stresses are not avoided, but employed actively, and the elastic deformation of the material becomes the parameter that gives the pavilion geometry its form.

The space-forming loadbearing structure of the pavilion was made from flat ribs of birch plywood 6.5mm thick × 10m long, arranged radially to form arches spanning 3.50m. The result was a torus with an outside diameter of 10m. In the longitudinal direction, the ribs were connected by tension-resistant "semi-recessed tenon joints" in a manner not unlike a zip fastener. In the transverse direction, each rib was interlocked with its neighbours at discrete points to form contrary waves of segments alternating in tension and bending; every tension segment maintained the form of the neighbouring bending segment elastically. So the intrinsic stresses in the bent ribs stabilised the structure.

Because in this structure the geometry, the loadbearing behaviour and the material are all mutually dependent, a design tool had to be developed in the form of a program that included the parameters which

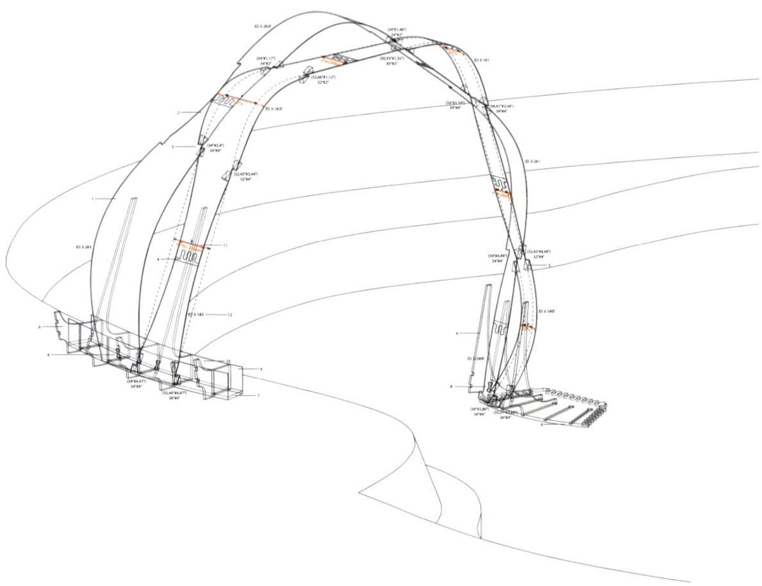

Each pair of arches consisted of segments alternating in tension and bending. Small timber wedges inserted at the interlocking points prevented the thin material from breaking at the wavy transitions.

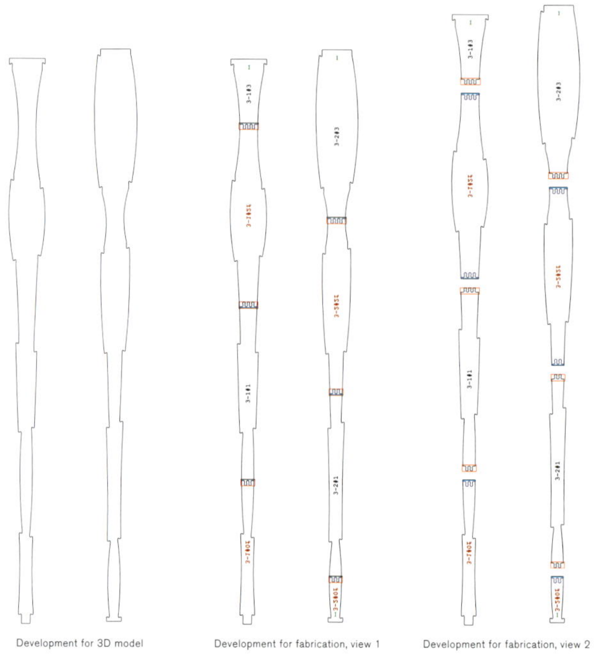

Rib geometries and segmentation for fabrication.

determine the form, e.g., the parameters specific to the material[4], which were determined in test series, and the values relevant for the structural analysis[5]. The outcome was a digital "information model"[6] that supplied the files for all the subsequent steps. Using the positions of the interlocks between the ribs, which were likewise stored in the information model, it was possible to check the feasibility of various pavilion geometries in the computer at an early stage of the design and generate these automatically as 3D geometry models.

The pavilion geometry finally built was made up of 80 individually shaped ribs in widths of up to 500mm, which were assembled from 400 segments; the separate parts were created by the different arcs and the offset interlocks. The cutting patterns for the ribs were calculated automatically by the data in the information model and supplied as CNC production data for fabrication on a six-axis CNC machine. Production and erection took just three weeks.

In that same year, the Laboratory for Timber Constructions (IBOIS) at EPFL, Lausanne, exhibited the results of its research into the efficiency and constructional aspects of timber structures at the "Timber Project" exhibition[7]. New wood-based materials and production methods were linked with digital methods of presentation and calculation for loadbearing structures in order to develop "new architectural forms in timber". One of the projects on show was the "textile module"[8]. Modelled on textile structures and weaving techniques, interlocking two double-curvature plywood boards

2 This pavilion was built in the summer of 2010 on the University of Stuttgart campus; it was the result of a joint research project by the Institute for Computational Design (ICD) and the Institute of Building Structures and Structural Design (ITKE). In order to investigate the relaxation behaviour of plywood and hence its suitability for loadbearing structures, the actual pavilion geometry was captured and recorded with a 3D laser scanner.

3 In bending-active structures, the form-finding is based on the elastic deformation of the material. Straight members or boards with a high elongation at failure (high load-carrying capacity but low stiffness) can be used to create complex, curving geometries; gridshells made from timber battens, built for the first time in the 1960s, e.g., the Multihalle in Mannheim by Frei Otto, are also bending-active structures. In those days the geometry was developed with the help of suspended models.

4 Experiments with models enabled the elastic bending behaviour of the plywood boards to be determined. Minimal bending radii, the ensuing stresses and how the deflection curves of neighbouring ribs depend on each other were ascertained and incorporated in the program.

5 The data found experimentally and the geometry developed from that enabled the writing of an FEM program that calculated the flexural and structural response while considering the actual behaviour of the material. The final dimensions of the ribs could therefore be determined taking into account the actual loads.

6 The "information model" was a specially programmed computer script that contained all the relevant data. One crucial task of this model was to manage the spatial positions and geometries of the interlocks between the ribs.

7 The exhibition took place from 26 February to 30 May 2010 in Lausanne under the direction of Prof. Yves Weinand, head of IBOIS.

8 Since 2007 the "textile module" has been the subject of investigation in the dissertation by Markus Hudert at IBOIS ("Timberfabric – Textile Assembly Principles and Wood Construction"). As the curvature of the boards – and hence the geometry of the loadbearing structure – is determined by the properties (elasticity and strength) of the material and the position of the fixing points at the intersections, the precise geometrical description of the basic model is very complicated and the subject of ongoing research.

resulted in an arch-type loadbearing structure from which it has been possible to produce different combination options for larger systems[9]. Adding together modules in the longitudinal direction results in "braided" arches that – placed in rows and joined together in the transverse direction with additional straight or wedge-type elements – results in a barrel-vault, "woven" structure. This planar weave-like configuration is intended to create a system effect, which in textile fabrics ensures that failure of one element does not have a negative influence on the load-carrying capacity of the overall construction.

The load-carrying capacity of the woven arch structure was tested using two prototypes of different sizes. The larger arch, about 2.10m high and spanning about 4.50m, was made from 40cm wide × 6mm thick plywood boards[10]. Loading tests revealed a fundamental and surprising feature: under vertical loading, the textile module becomes stable at the centre of the arch – the bent plywood boards extend in the longitudinal direction, and the triangular cross-section[11], which is responsible for the load-carrying capacity, becomes narrower and deeper.

The experimental structures of the University of Stuttgart and IBOIS are impressive and surprising owing to their unusual structural forms that broaden the spectrum of efficient, bending-active structures. At the same time, a design tool that allows the drawing to be a direct image of the reality opens up new opportunities for architecture.

Besides the development of new, bending-active structures in timber, digital design tools and production methods are also leading to a new understanding of and new look for wall and loadbearing structures. Under the heading "Sequential Structures", the students at ETH Zurich are investigating the constructional and

The loadbearing structure was raised to form an entrance to the pavilion. In this area the ribs were extended to form narrow columns. Experimental pavilion, Stuttgart, Germany; Institute for Computational Design (ICD) and Institute of Building Structures & Structural Design (ITKE), University of Stuttgart, 2010.

architectural potential of walls and loadbearing structures created through the addition of small timber members[12]. With the help of a robot-controlled production plant, each member is individually cut to its specific size and positioned exactly within the total structure. This method results in relief-type, curving walls that change continuously and subtly – flat surfaces merge seamlessly into curved surfaces or projections and recesses develop continuously along curving lines. The walls are therefore given a modelled, sculptural character and the semantic level of the individual modules shifts from the constructional to the ornamental. The use of digitally controlled robot production results in individual forms for the module and the overriding (wall) system, too – both formerly standardised elements and components in terms of their geometry.

A temporary pavilion demonstrated the artistic potential and load-carrying capacity of sequential timber

9 The research work investigated potential timber structures that can be developed by adding "textile modules" together (different structures composed of "double-layered build-ups" are the outcome of the research to date).

10 The second arch, about 60cm high and spanning about 1.60m, was made from 10cm wide × 2mm thick plywood strips.

11 At the centre of the arch, the plywood boards form two sides of an open triangle in cross-section. Under vertical load, the angle between the sides of the triangle decreases, thus increasing the load-carrying capacity.

12 Parallel with and complementary to the work of the students, the department works with additive design and production processes. The focus of the work is the relationship between design and production processes. The theory postulated was that the use of digital design and production methods constitutes a new digital "trade". The opportunity to process large amounts of information is bringing about a qualitative change in architecture. The intention behind observing additive processes using an industrial robot is a conceptual system in which the configuration of individual production processes represents an integrative component in architectural design. Tobias Bonwetsch is working on a dissertation on this subject at the department.

The "woven" arches at the "Timber Project" exhibition (since 2010) consisted of inner and outer arch elements such that the interstices were covered. Adding together "textile modules" in the longitudinal direction was achieved with two timber members introduced between the plywood strips of the respective module. Timber wedges between the module layers constituted the transverse connections; IBOIS laboratory at EPFL.

structures. Sixteen columns with different geometries were built from layers of timber slats according to a defined construction principle in order to create a covered, open structure. Projections, returns and rotations of the individual layers of timber resulted in columns with complex three-dimensional geometries. Starting with a square base measuring 0.40 × 0.40, the column shaft tapered and then widened constantly and smoothly before terminating in a 2.00m square column head. Although these boundary conditions were valid for all 16 columns, rotating the bases of the columns through different angles with respect to the column heads led to twisted column surfaces with different geometries. Calculating the positions of the timber slats and their precise assembly was carried out by the industrial robot. The timber modules were connected by bolts and tie-down straps as used on goods vehicles. The straps, placed within the hollow columns, tied together the top and bottom of each hollow column and counteracted the tensile stresses in the discrete connections[13].

A further development of the prefabricated, additive structures was developed within the scope of the In-Situ Robotic Fabrication research project[14]. The main focus of this project was the adaptability, flexibility and mobility of digital production as well as the interaction of people and machines during the construction process. The work resulted in a mobile robot unit[15] that can help to transfer the production process to the building site. As a prototype, a "fragile wall structure" in the multi-storey car park at ETH Zurich was built from timber modules. The mobile robot unit, which positions itself to suit the circumstances, assembled 1000 modules to form a curving, room-high wall. In order to be able to react to the constantly changing conditions of the construction process and be able to identify components and dimensional tolerances, the robot is equipped with a laser system that signals the given circumstances to the robot. The measurement data is integrated into the construction strategy and the position of each element calculated individually.

The robustness and adaptability of the mobile production unit were demonstrated at the "Fabricate 2011" conference in London, where an architecturally indeterminate wall structure was fabricated and erected using timber modules of different heights (30, 45, 60mm). The robot reacted to unevenness and inaccuracies in the modules positioned beforehand (a situation exacerbated by the different module heights) by adjusting the drop height and placing rate and correcting the module position without interfering with the given geometrical structure.

The geometry of additive wall structures is determined by defined rules that are programmed directly according to the construction logic. The choice of material and module size leads to specific conditions for the assembly process (overhangs, stability during construction, loadbearing capacity, overall stability, etc.) which determine the "design" depending on the production options. Decisions according to aesthetic/artistic aspects are checked empirically on the model within the given set of rules. The construction of additive structures by mobile robots combines production and erection and thus demonstrates a next step in the evolution of intelligent architectural production processes.

Besides the projects prompted by materials or production aspects, more and more projects inspired by

13 The ties were connected to the steel plate at the base of each column and steel cross-beams at the top; their primary purpose was to stabilise the structure during transport and erection.

14 The In-Situ Robotic Fabrication (dimRob) project was part of an international, EU-sponsored project known as "ECHORD – European Clearing House for Open Robotics Development". Headed by Prof. Alois Knoll from the Department of Robotics and Embedded Systems at Munich's science and technology university, the project promotes collaboration between academic research and industry in the field of robot technology.

15 The new robot was developed at ETH Zurich in conjunction with Bachmann Engineering AG and presented at FABRICATE in London, 2011.

Sequential wall structures are the outcome of the addition of small timber sections; Gramazio & Kohler, ETH Zurich.

bionics have been seen in recent years. Researchers and architects have been experimenting with honeycomb, folded-plate or "performative" structures in an attempt to transfer the efficient structures of nature to architecture. Honeycomb structures, which in nature are found in beehives, cell arrangements or tortoise shells, are particularly suitable for creating irregular forms. Designed as a grillage, they represent a reinterpretation of the traditional coffered or waffle slab. Whereas the individual panels of such a slab are normally formed by a grid of beams or ribs, the honeycomb grillage is generally assembled from glued laminated timber elements joined together to form boxes and connected by steel gusset plates and steel dowels. In a simi-

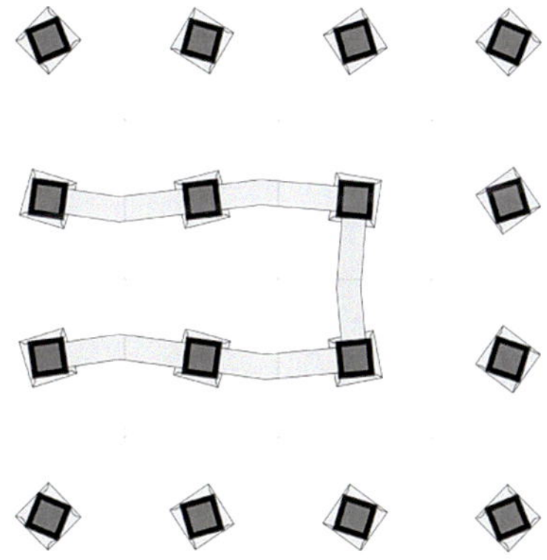

Plan "West Fest Pavilion".

"West Fest Pavilion" was used as a bar during a three-day festival to inaugurate a new section of motorway. The timber slats (60 × 40mm) of the prefabricated columns were individually fabricated and positioned by a industrial robot; Gramazio & Kohler, ETH Zurich, 2009.

lar way to cell structures in nature, the irregular geometries of each neighbouring honeycomb are mutually dependent[16]. The irregularity of the honeycomb forms resulted in a multitude of different member lengths and intersections at the nodes, which together led to a low repetition rate for the member geometries. However, the introduction of digital CNC production in timber construction allows such grillages to be readily and economically produced from quasi-bespoke individual pieces. Coupling digital design and production with the aim of a digital chain from the first conceptual sketches right up to production was the focus of the design for a pavilion at the 2011 National Horticultural Show in Koblenz[17]. The polygonal grillage, reminiscent of a covering of foliage, spreads out from the shell-like timber columns so the loadbearing structure deforms three-dimensionally. Using a 3D geometry model, the grillage and its deformations could be ascertained, designed and fabricated. Employing new technologies that depart from rigid, right-angled grids helps the loadbearing structure itself become a decorative feature.

Folded foliage arrangements or origami methods demonstrate impressively the stabilising effect of the folds and the efficiency of folded plates[18]. With the production of large-format wooden boards, knowledge of the material characteristics and the introduction of digital design, calculation and production[19] – which enabled the tessellation of random bodies, the determination of internal forces and the uncomplicated fabrication of individual parts with different forms – timber folded plates are again being noticed by architects and researchers[20]. However, whereas paper or thin metal sheets can be folded or bent without any problems, timber folded plates have to be made from individual boards joined together, with structurally effective connections at the "folds". The constructional solution for transferring forces at these narrow lines of contact is the main problem that has to be solved when building folded plates from flat timber boards[21]. Traditional connections using timber gusset plates, punched metal fasteners and steel fishplates usually upset the lightness and simplicity of the folded structures and the "look" of the folds. However, gluing or screwing the edges together to create an "invisible" connection is currently only possible for folded plates made from glued laminated timber elements of sufficient thickness.

In order to deal with this issue, RWTH Aachen University has developed folded plates with a new type of edge connection, the "textile joint", as part of a research project[22]. The basis and raw material for the "textile joint" are two laminated veneer lumber (LVL) boards either side of a layer of polyester fabric glued in place with resorcinol resin. At the folded edge, a slit is cut in this composite board material on one side and on the other side it is machined to a V-shape so that at the edge the boards are connected exclusively by the sandwiched textile, which means that true rotational movements and folding processes are possible. Various folded loadbearing structures made from textile-timber composite boards measuring 1.80 × 27m, specially produced for the research project, were developed and tested during the work. The results include arch-shaped structures with triangular, opposing folds, double-curvature structures made from individually shaped, hexa-

Demonstration at FABRICATE in 2011: the robot built up 1330 timber modules, 250 × 100mm with varying heights, to form an architecturally indeterminate wall structure.

16 In 2005 Ludger Hovestadt and his students at the Department of Computer Aided Architectural Design (CAAD), ETH Zurich, carried out experiments involving box-type "cell structures" made from particleboard, which they connected to form a spherical pavilion 4m in diameter. The pavilion, which was developed for the SWISSBAU 05 fair, demonstrated the possibilities of a digital chain from design to production. The irregularity of the "cells" resulted from the digital simulation of "disrupted growth processes" – cell growth was simulated around given square openings.

17 The "treehugger" pavilion was the outcome of a joint research project by the Koblenz Chamber of Trade, the Digital Design Department of Trier University of Applied Sciences and Düsseldorf-based architectural practice one fine day. Various digital design tools that link preliminary and detailed design work were checked by means of the pavilion concept.

18 Folds can be essentially divided into two categories: longitudinal folds with continuous edges and alternating hips and valleys in parallel, curved or skewed arrangements, and point or facet folds whose edges meet at a point to create faceted triangular or square surfaces for tetrahedron- or pyramid-type folded arrangements.

19 Prior to the introduction of digital design and calculation methods, it was only possible to handle regular geometric bodies with regular folds which could be described mathematically (parallel longitudinal folds over a rectangular plan form or longitudinal folds meeting at one point over a circular/square plan form).

20 The 1950s and 1960s were the heyday of folded plates. Delicate folded plates in reinforced concrete designed by Pier Luigi Nervi and Félix Candela created a furore. Renzo Piano, Arthur Quarmby and other architects experimented with folded plates made from polymer sheets. A number of remarkable folded-plate structures in timber appeared worldwide, too. However, as neither the board materials of the time nor the calculation methods were suitable for building folded plates from single wooden boards, the timber folded plates were made up of linear elements partially covered with boards. The only exceptions to this are the Annette Kahn residential home in Tübingen, Germany, a folded plate with opposing longitudinal folds which was built in 1962 from 6cm Wolff double-wall panels, and a folded-plate roof for the IBA 1976 fair in East Berlin, which was made from two layers of boards arranged in a diagonal pattern (see the dissertation by Katharina Leitner, 2004).

21 The connections along the narrow lines of contact must be able to transfer vertical and horizontal hinge forces plus shear forces.

22 Prof. Wilfried Führer (Chair of Structures and Structural Design, RWTH Aachen University) has been investigating plate and shell structures made from wood-based board products together with the Kerto and Finnforest Merk companies since the 1990s. Katharina Leitner's dissertation on loadbearing structures made from board-type wood-based materials with textile joints investigates the possibilities of forming folds with the help of embedded textile materials. She has also supervised student seminar papers.

The vaulted roof of pentagonal frames made from 8cm thick × 40cm deep glulam is supported on five internal columns with a shell-like, curving form that creates a fluid transition to the roof structure itself. The curvature of the grillage leads to a "shell effect", thus obviating the need for a perimeter beam. "Treehugger" pavilion, 2011 National Horticultural Show, Koblenz, Germany; Trier University of Applied Sciences (Prof. Hoffmann) with Koblenz Chamber of Trade and architectural practice one fine day.

gonal honeycombs, a tessellated circular lattice dome and an arch-shaped spatial structure with spans of 2.50m[23]. Irrespective of the efficiency of the structures, the simplicity of the principle of the "textile joint" itself was impressive enough: the slit, large-format boards could be folded with ease like a sheet of paper without the need for any elaborate connection details[24].

Likewise unsolved, or at least very difficult, so far is the determination of the loadbearing behaviour of folded plates, which is explained by the multitude of influential factors but also the lack of adequate design tools. At IBOIS in Lausanne, researchers are getting closer to solving this problem and have developed a computer program that can readily illustrate, change and check the design of complex folded plates in a 3D model; it can also be easily linked to structural design software[25]. The computer program represents a folded plate by means of two polygonal lines: the fluting profile, which determines the height and angle of the main folds and essentially defines the load-carrying capacity of the folded plate, and the cross-sectional profile, which describes the folds transverse to the principal folds. The program defines the angle of the reverse folds and hence determines the geometry of the folded plate. Using both lines, the program can generate the folded plate so that the architectural design and structural analysis of different types of folded plate can be checked quickly or the desired type of folded plate adjusted and defined to suit the structural requirements. Coupled with digitally controlled CNC plant, the geometrical data of the folded-plate surfaces can be input directly and the boards fabricated with ease.

The new St. Loup Chapel in Pompaples, Switzerland (see p. 55), provided an opportunity to use the new method of presentation in a real project for the first time. The folded plate for this chapel consists of angled longitudinal folds with parallel valleys and angled hips in the roof surface which ensue due to the differing and alternating opposing amplitudes of the main folds along the plan lines of the two longitudinal walls. In addition, the folds, both on plan and in elevation, are arranged on a curve and therefore form different angles and heights. The folds of the fluting and cross-sectional profiles are mutually dependent (modulation on one line

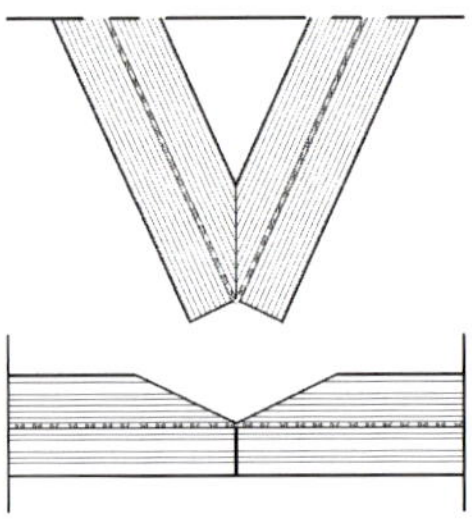

Textile joint. Katharina Leitner, RWTH Aachen University, Finnforest Merk, 2000–2003.

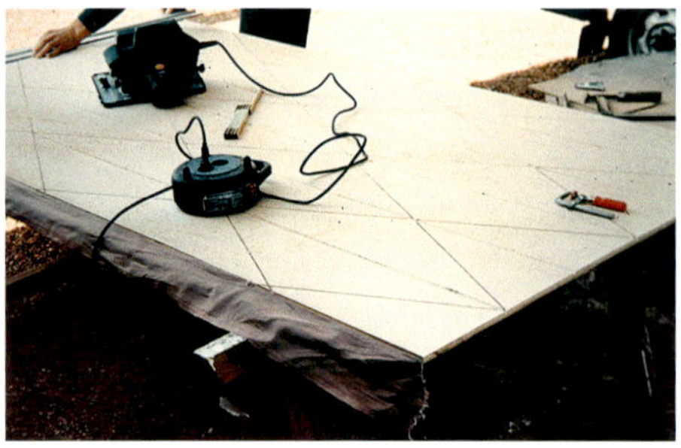

The basis for producing the folded plates was a composite material with layers of LVL either side of a textile core specially produced for the experiments.

Honeycomb structure: The honeycombs are made up of diamond-shaped boards connected together by the textile material and fixed in the desired position with steel angles (design: C. Grosse Kathöfer, M. Soldan). Folded arch structure: The angles of the routing determine the exact orientation of the folded-plate surfaces with respect to each other. The tessellated dome is made up of 120 triangular segments created by folding 24 strip-type plywood panels (2 No. 6 mm); design: L. Berger, R. Albrecht, A. Scheibe, J. Voelker, J. Vielhaber.

has an immediate effect on the other line) and so the continuous height development of the volume is a result of the curved, bulbous outline on plan.

Modulations of the originally regular, "simple" folded plate led to individual cutting geometries for the folded-plate surfaces, which for fabrication are taken directly from the design program. The 3–7m high folded plate is made from 40 and 60mm thick cross-laminated timber (CLT) boards screwed to bent perforated plates along the folds and connected together at the base by ties that prevent the chapel walls from "spreading apart"[26]. With roof panels just 60mm thick spanning up to 9m, this chapel is an impressive illustration of the efficiency of folded plates. The concertina look of the folds and the ensuing changes in height inside, with the highest part above the altar, plus the rhythmic effect of the interior due to the folds turn this folded plate into a modern interpretation of church architecture.

The fascination of folded plates, besides their structural efficiency, is essentially due to their spatial effect, which is primarily determined by the pattern of folds. The smaller the individual surfaces are, the more

23 A doctoral thesis was submitted to RWTH Aachen University, Chair of Building Construction 1 (Prof. W. Führer). The experimental structures were built in the years 2000–2003. Owing to the lack of data on the material and suitable FEM element types, the calculations for the folded plates (determination of internal forces) could only be approximated. In order to put the load-carrying capacity of the textile connection to the test on a realistic scale, a folded arch segment with triangular surfaces was constructed from 27mm thick LVL boards. With a span of 7m and height of 3.50m, the assembly illustrated the potential for folded plates. A mistake led to a non-loadbearing textile material being glued in place and so the individual boards of the folded plate were connected by transparent plastic fishplates (Makrolon®) screwed in place.

24 The embedded textile material transfers shear forces and their tensile force component. Compression forces are transferred by contact between the timber boards.

25 Within the scope of his doctoral thesis "Origami – structure plissée en panneaux de bois massif" at IBOIS, Hani Buri studied the potential of folded structures for architecture. He developed the new, digital method of displaying folded plates as part of this work. In order to verify the method and also the structural values he worked out, he built a folded arch structure spanning 6.70m from 21mm thick, trapezoidal, spruce plywood elements arranged in a herringbone pattern, which were connected with 80mm long screws. The deformations of the loaded structure exceeded the values calculated and the connections failed at a load of just 27kN.

26 Despite the rigid screwed connections along the folded edges, which cause local fixity moments at the folds, these edges were considered as hinges in the structural calculations.

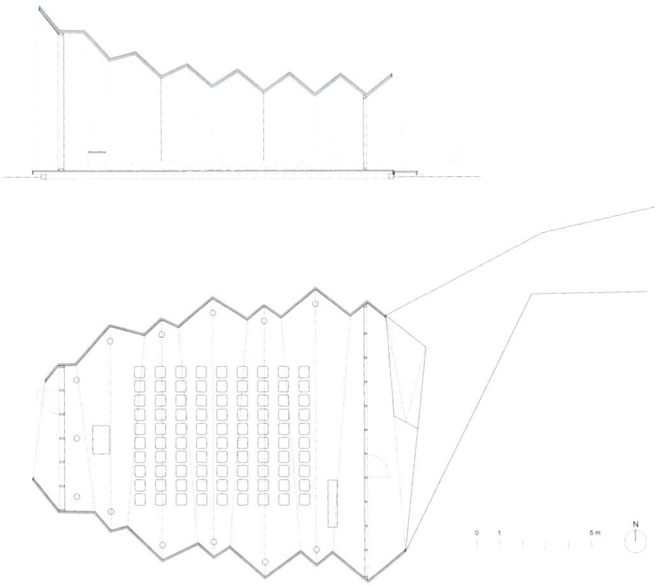

Plan of and section through the chapel in Pompaples.

Impregnated 3-ply core plywood (d = 19mm) was used for the external envelope, fixed to the structure with timber battens and open joints above the bituminous water run-off layer. The end façades are made from polycarbonate double-walled sheets.

complex and atmospherically denser are the sculpted spaces inside the building. However, the folding problem increases with the compartmentalisation and the number of separate surfaces. Push-fit connections for an arch-type structure made from folded timber modules represent a new type of jointing tried out at IBOIS. The folded timber modules consisted of trapezoidal 3-ply boards. The alternating folding along the short or opposite long side of the trapezoidal surface results in two different modules which, like a house of cards, are inserted into each other in turn. To achieve this, each module has diagonal slits top and bottom, the angle of which changes continuously with respect to the plane and section of the board so that fitting the modules together produces the desired arch shape. As the legs of the modules intersect, the result is a three-dimensional loadbearing structure with a different appearance on the inside and outside of the arch.

Another structure built from three-dimensional modules was the amorphous, cave-like timber shell made by the University of Stuttgart, which combined honeycomb and folded-plate concepts. This pavilion was the result of research concerned with how biological structure-formation processes can be transferred to architecture by means of new digital design and production methods. The work involved studying the spherical shells of sea urchins and using the structure-forming features as parameters for the design process in a design and simulation program specially developed for this purpose. Combining this with the material-specific properties of plywood and the loadbearing behaviour of the connections, which were determined experimentally, enabled the "morphology" of the pavilion to be developed and optimised on the computer[27].

The result of these "morphological processes" was a shell structure[28] made from folded, honeycomb-type modules, which are in turn assembled from glued plywood pieces. As with the plate skeletons, three elements meet at every point in both the higher system (the modules) and the lower system (the panels). Similar to the living organism, which reacts to different conditions during its growth process by adapting its geometry, the modules have irregular, individual forms to suit the flow of forces and the pavilion geometry. In those areas of the shell with less curvature, the honeycombs are about 2m across, whereas those at the edges, the openings and in areas with greater curvature are only about 0.5m across. The irregularity of the structure continues in the geometry of the box joints, which are different at every joint between panels. Only through digital precision work and by adding another

The load-carrying capacity of the structure was tested on a prototype with a span of 7.50m and a rise of 3.50m at the EPFL campus. The folded modules made from 21mm thick 3-ply boards were fabricated on a five-axis CNC machine because the slits are angled with respect to both the plane and the section of the board.

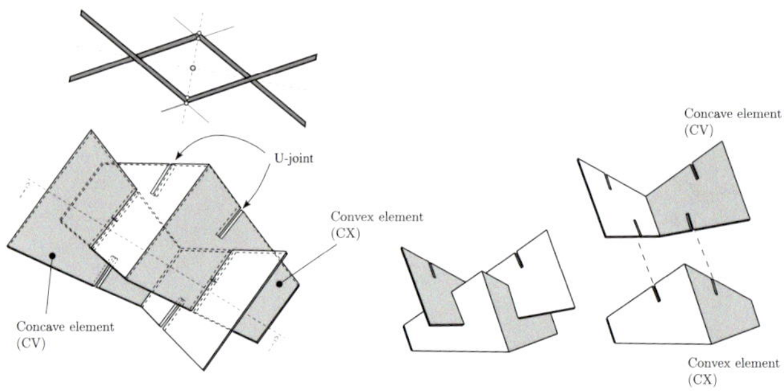

The 3-ply boards (side length = 80cm) were connected by glued-in dowels and screws. The direction of the grain of the core ply was transverse to the screw direction and therefore guaranteed a loadbearing connection; Prof. Yves Weinand, IBOIS, EPFL.

About 50 different, three-dimensional modules were joined together to create a timber shell about 70m² in area and 4.50m high. Experimental pavilion, Stuttgart, Germany; Institute for Computational Design (ICD) and Institute of Building Structures & Structural Design (ITKE), University of Stuttgart, 2011.

27 As the structure-forming and structural analysis parameters were integrated into the design program and there was a constant link back to the finite element software, the angles between the elements and the flow of the forces could be checked and modified at any time.

28 The non-rigid connections (glued box joints + screws) between the elements result in a ductile shell capable of accommodating flexural stresses.

The spherical plate skeleton of the sea urchin, which is composed of polygonal, three-dimensional modules interlocked along their edges, was the model for the design of the pavilion. The 6.5mm thick, trapezoidal or hexagonal plywood panels meet at angles varying between 15° and 165° and are connected by glued box joints. Both the design and the fabrication of the box joints, positioned arbitrarily in space, represented a challenge.

The pavilion contains a main room and a secondary room. The latter is created by separating the internal and external skins of the hollow body so that it is possible to walk through the inside of the modules, so to speak. Fixing the pavilion to a ground slab secured it against wind uplift.

machining axis to the six-axis CNC router[29] was it possible to fabricate exactly the 850 individually shaped panels with their different box joint geometries.

The digital simulation of the laws governing organic structures enabled a design method to be devised which allows any type of geometry to be formed from irregular modules.

Another example for the transfer of natural processes to architecture is the self-supporting, reactive wall system that alters its shape to respond to changing microclimatic relationships[30]. The model for this was the responsive, reversible mechanism of a spruce cone, whose scale-like structure opens or closes depending on the humidity. The interactions due to environmental influences, material properties and system behaviour, as observed in the spruce cone, constitute the basis for the wall system. Instead of counteracting the normally undesirable deformations due to shrinkage and swelling, the natural properties of the wood are exploited in order to control the wall system's permeability to light and air. Series of tests revealed the deformations and reaction times of veneer composite elements exposed to changing hygroclimatic conditions and depending on the direction of the grain, the adsorption or resorption capacity and the ratio between panel thickness and panel size. Based on this, it was possible to develop diamond-shaped modules made from a loadbearing framework and functional layers of moisture-sensitive veneers. Adding these together results in a curving, free-form wall surface made from loosely overlapped wood shingles, which join up as the humidity increases due to differential stresses in the material and make the surface rainproof. This reaction is reversible and is based on constant material properties.

In order that the responsive structure can be used under real conditions and the opening and closing of the "skin" functions according to the requirements, the moisture content of the veneers must be matched to the climactic conditions of the surroundings at the time of building the structure. When erecting the structure on a sunny day, the moisture content of the curving, open structure matches the relative humidity of the surroundings, which means that when it rains, and the humidity rises, the fibres increase in length and close the structure to make it rainproof[31].

But beyond a culture of building aimed at efficiency, various universities, through interdisciplinary cooperation and linking digital design and production tools, have in recent years built experimental structures that demonstrate new paths forward for timber construction. Wood is stacked, braided, bent, woven or folded. The ensuing structures expand the spectrum of timber loadbearing construction and represent an aesthetic enrichment for architecture.

Responsive wall surface: the diamond-shaped "timber scales", each about 0.80 × 0.80 m, open or close depending on the humidity of the air; Prof. Achim Menges, Steffen Reichert, Institute for Computational Design (ICD), University of Stuttgart, 2010.

29 Placing the panels on a turntable for machining and coupling this with a six-axis robot resulted in a seven-axis machining system.

30 The "responsive surface structure" was developed in 2005–2007 by Steffen Reichert at the Offenbach University of Art and Design, Department of Form Generation and Materialisation, and supervised by Achim Menges. As a follow-up to their work, Menges and Reichert are investigating the application options of hygroscopically actuated veneer composite elements as part of the "Biomimetic Responsive Surface Structures" project at the University of Stuttgart, Institute for Computational Design. Together with the architectural practice Scheffler+Partner, the Institute developed a pavilion in 2010 which has a free-form, honeycomb-type timber loadbearing structure that closes during rainfall owing to the material properties of the veneer composite.

31 Experimentation led to the development of outer layers of veneer and synthetic intermediate layers with an optimum fibre saturation point so that the angle of the opening can be adapted to suit the local climatic conditions.

Projects

Trade fair hall 11

Frankfurt am Main, Germany

Location	Messegelände Frankfurt am Main, Ludwig-Erhard-Anlage 1, 60327 Frankfurt am Main, Germany
Client	Messe Frankfurt Venue, Frankfurt am Main
Construction	2007–2009; completed: July 2009
Architects	Hascher Jehle Architektur, Berlin
Project management	FAAG Technik GmbH, Frankfurt am Main
Fabrication drawings	ATP Achhammer-Tritthart & Partner, Munich
Structural engineers	RSP Remmel + Sattler, Frankfurt am Main
Timber construction	WIEHAG, Altheim
Main timber product used	glued laminated timber (spruce)

Khaled Saleh Pascha

The two-storey hall 11 on Frankfurt's trade fair grounds. The roof with its long cantilevers, all in glued laminated timber, appears to float above the reinforced concrete blocks.

Together with the new entrance building, this new trade fair hall, finished in 2009, rounds off the western end of Frankfurt's trade fair grounds in their urban setting. The roof structure to the hall is one of the most impressive examples of the use of timber for long-span designs.

This complex, comprising a two-storey trade fair hall and a new west entrance facility, the result of a competition held in 2006, boasts impressive dimensions: 196.7 × 114.8m overall, a height exceeding 27m and clear internal hall dimensions of 156 × 76m, which make hall 11 one of the largest exhibition buildings in Frankfurt. With more than 11 900m² of space on each of the two floors, it is exceeded only by halls 3, 4, 8 and 9. Whereas the ground floor has intervening columns, on the upper floor the roof spans 79m over uninterrupted exhibition space.

In the competition submission, the roof was in structural steelwork. However, during the detailed design phase, the high steel prices at the time of the tender led to steel being replaced by timber. The alternative timber design proved to be considerably less costly, resulting in savings of about €5 million. The timber roof was also 100t lighter (1250t instead of 1350t in steel). Apart from that, the use of timber, a renewable material, sequestered some 2160t CO_2, whereas the production of the steel alternative would have released 1740t CO_2.

The roof to hall 11 is a simply supported structure with a span of 79m and a 17.40m cantilever on each side. These long cantilevers lend the building an elegant, dematerialised appearance.

The glued laminated timber trusses (in spruce) are max. 6.60m deep and have round steel bars as diagonals. In contrast to timber diagonals, this results in a very delicate design, which in turn leads to the entire construction appearing very transparent and lightweight despite the huge dimensions. Each diagonal is made up of one or two steel ties with cross-sections to match the actual loads. The diagonals descending from the support point are designed for dead loads, the prevailing load case, and consist of two parallel round steel bars in each bay. The diagonals in the opposite direction are required only for the uplift load case, which can occur in the cantilevers or in the main span; accordingly, only one smaller tie is needed in each bay.

The long cantilevers include diagonals made from glued laminated timber, which function as ties and struts. However, these diagonals are hidden behind the cladding to the underside of the cantilever.

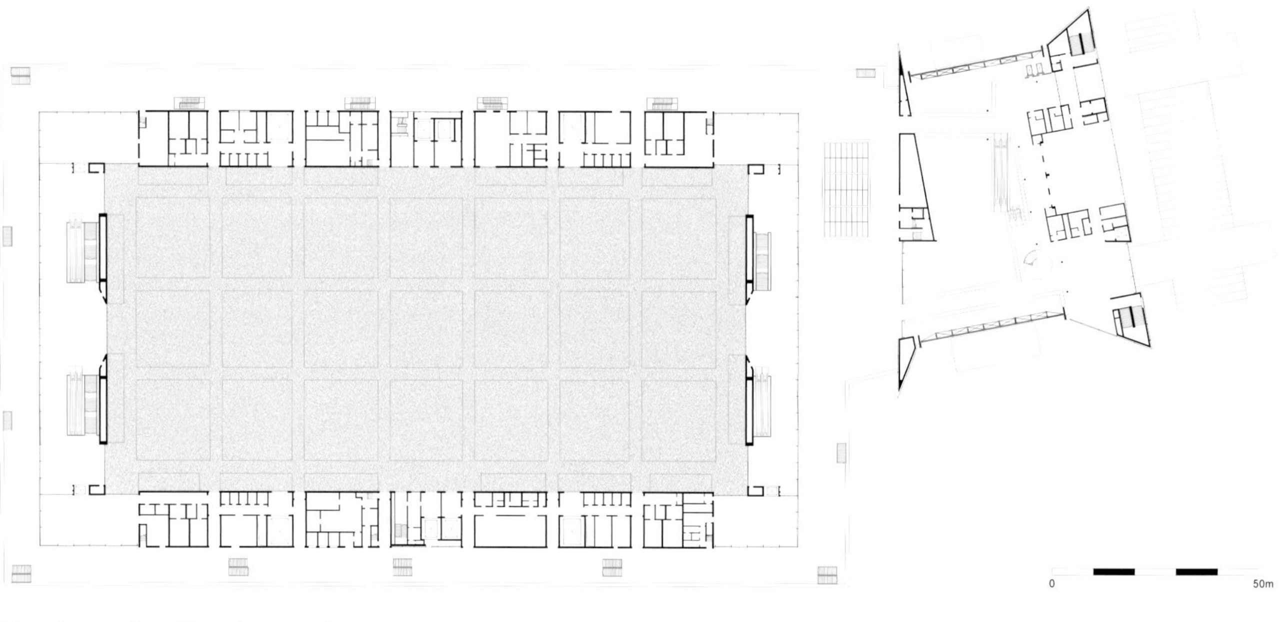

Plan of upper floor. The primary roof trusses are supported on the inside walls to the ancillary facilities blocks. The new entrance building is shown to the right of the hall. This is not only an entrance to hall 11, but also a new main entrance for the trade fair grounds and represents part of the overall planning of the Hascher Jehle architectural practice.

The truss during construction (the outline of a person is included to give an idea of scale). Each diagonal consists of one or two steel ties with cross-sections to match the actual loads.

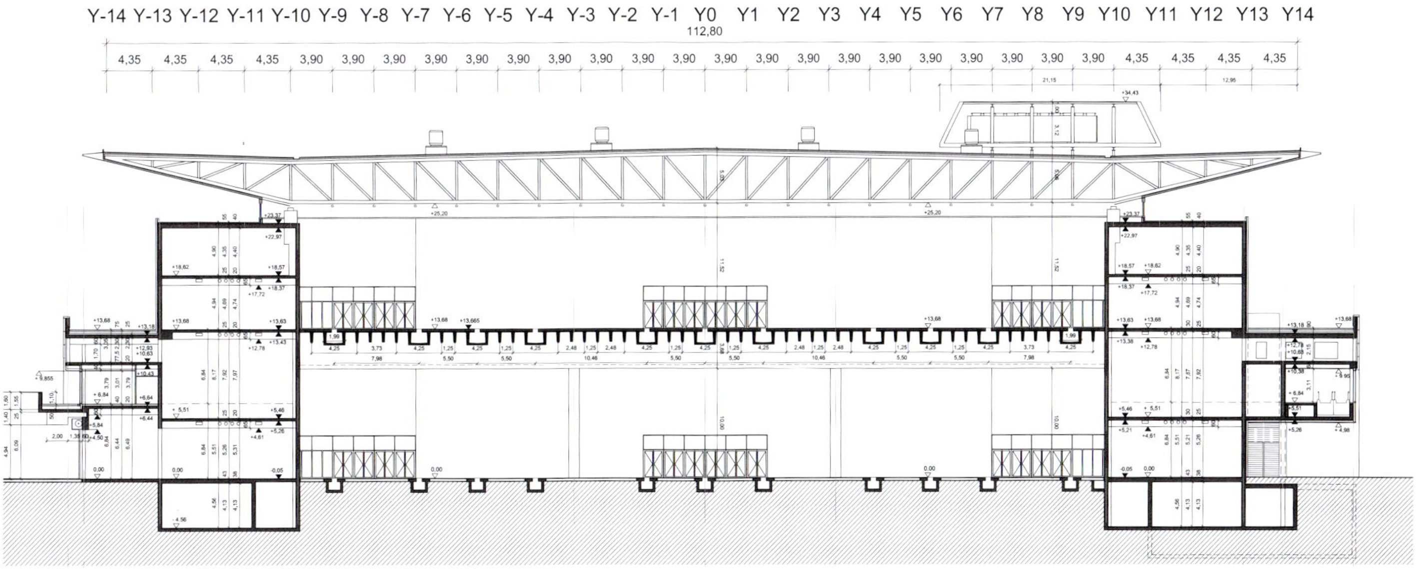

Section through the two-storey hall. Recooling units are mounted on the roof. A row of smoke vents in the roof ensures that smoke and fumes can escape in the event of a fire.

Timber secondary beams link the top chords of the trusses, which are at a spacing of 10.40m. As the roof cantilevers on all sides of the building, the secondary beams require adequate structural depth to handle the considerable bending stresses due to the cantilever. Therefore, in the two end bays at each end of the building, every fourth secondary beam is a truss with a depth the same as the primary truss.

The entire roof structure is supported without restraint at 38 points on the reinforced concrete walls to the blocks housing ancillary facilities, which run the full length of the building. Reinforced elastomer bearings ensure that changes in length due to temperature fluctuations or other loads do not transfer any additional shear forces to the walls.

With such long cantilevers on all four sides, securing the roof against uplift is critical. Therefore, additional steel cables that are activated in the event of wind uplift forces run parallel with the vertical timber posts at the supports.

The enormous dimensions of the truss chords rendered full prefabrication of the truss impossible. Assembly on site was therefore carried out on trestles, with each truss lying on its side. Each top chord was prefabricated in two 39m long segments and assembled on site. Three segments were necessary for the bottom chord, the middle one of which was 50m long. The connection for transferring the tensile forces consists of steel splice plates and diagonal bolts.

Grade GL 32c timber to DIN 1052 was used for the truss chords (GL = glued laminated timber, 32 = permissible bending stress in N/mm^2).

The colossal forces that occur at the interfaces between the trusses, and a number of particular geometrical issues in this design, led to several new developments being necessary for the connections. For example, full-thread screws are inserted on both sides of the steel plate, which are at 90° to the steel dowels and perpendicular to the direction of the grain of the wood. It was proved in tests that this strengthening results in

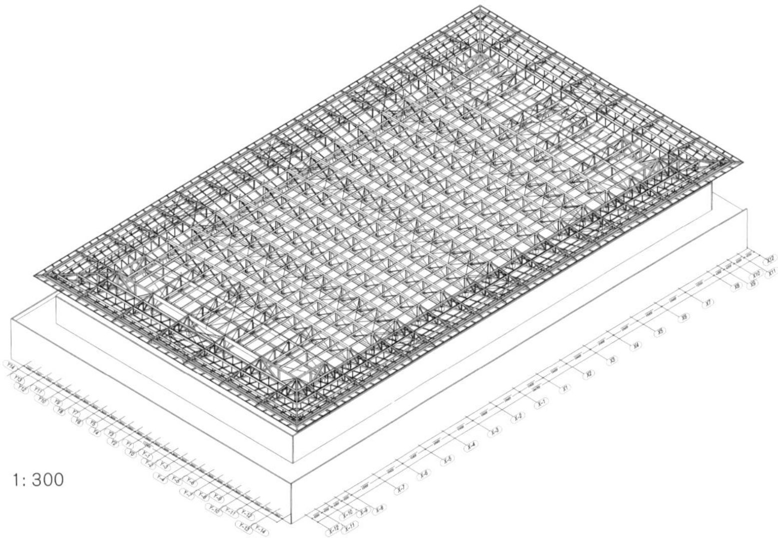

1: 300

Isometric view of roof structure. The secondary beams in the form of trusses can be seen in the end bays of the building. Trusses are necessary here to support the 17.60m cantilever of the roof at both ends of the building.

an ultimate load at the node that is 380% higher than without the strengthening.

The roof structure has been left exposed internally, i.e., the trusses are neither clad nor concealed behind a suspended ceiling. The only treatment given to the structure was a coat of glaze on the underside.

Hall 11 is a demonstration of the dimensions and efficiencies possible with modern long-span timber structures. Just a few decades ago, structures that were seemingly only feasible in steel, reinforced concrete or possibly in the form of expensive membrane designs –

The outer end of the cantilevering truss during construction. Steel plates let into the timber and steel dowels are used for all connections. Full-thread screws in the opposite direction to the grain of the wood prevent splitting at the connections.

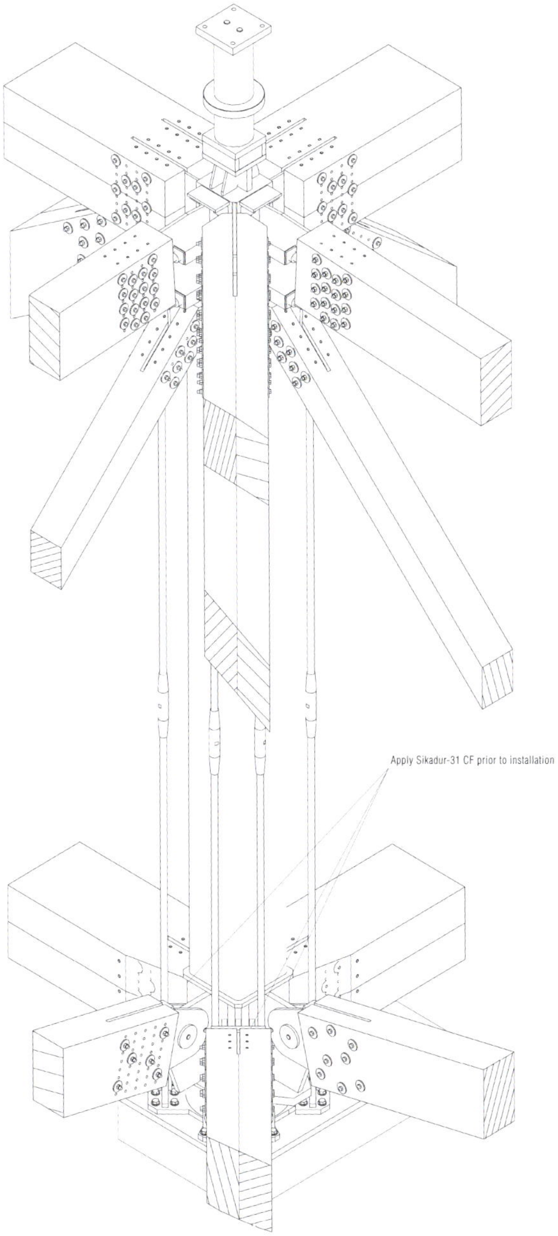

Detail of outer corner of roof structure where cantilevering roof is connected (chords with smaller dimensions). The all-steel connection at the junction between primary truss and cantilevering truss can be seen, likewise the steel cable ties to prevent uplift at the corner post of the truss.

Steel plate connection on top chord of truss at corner of roof. The projecting steel plates connect to the top chord and diagonal of the truss joining at a right-angle. The downward extension with the single hole will later be connected to a vertical steel cable, which is in turn connected to the support to resist uplift.

11 930m² of uninterrupted floor space beneath a 79m span and a 17.40m roof cantilever on all sides – can now be built in timber at a competitive price. Even fire protection measures, which can hamper timber construction on occasions, did not constitute a serious problem here; the skilful positioning of relatively shallow smoke reservoir screens and exhaust ducts in the roof satisfied the fire protection requirements. It was therefore possible to leave the roof structure exposed internally without the need for any further fire protection measures. And building the roof in timber instead of conventional structural steelwork also results in a CO_2 balance that is nearly 4000t better, which shows quite impressively that timber is not only equivalent to other construction materials, but in some ways far superior.

Detail of connection for diagonal steel tie. The holes are predrilled to suit the angle of the full-thread screws.

The exposed roof structure over the upper floor of the trade fair hall.

Business premises, BIP Computer

Santiago de Chile, Chile

Location	Avenida Francisco Bilbao 2296 – Providencia, Santiago de Chile, Chile
Construction	2006–2007; completed: 2007
Architects	Alberto Mozó Studio www.owa.cl
Project team	Alberto Mozó Leverington, Francisca Cifüntes, Mauricio Leal
Structural engineers	Juan López Ingenieros
Contractor	Estructura Madera Arauco S.A + Constructora Las Torcasas
Main timber product used	glued laminated timber (Monterey pine)

Khaled Saleh Pascha

West elevation of BIP building.

The business premises for BIP Computer are located in the Providencia district to the east of the centre of Santiago de Chile. The new building has retail space for computer sales on the ground floor and company offices on the upper floors. Two existing buildings, dating from 1936, were integrated into the new plans to accommodate further retail space, a storage area and a workshop. One fundamental requirement the architect had to comply with was the quick and easy erection and dismantling of the structure because the building is intended only as an interim use of the plot.

Architect Alberto Mozó developed a loadbearing system based on standard sections (Hilam Estándar) with dimensions of 342 × 90mm supplied by the Chilean timber fabricator Arauco. The glued laminated timber frames can be produced in lengths of up to 30m, although in this case the standard length of 10m was used for the diagonal columns and horizontal beams. One key advantage of adapting column and beam dimensions to the sizes of standard cross-sections was the considerable cost savings: the cubic metre price of glued laminated timber in these standard dimensions, fabricated and ready to install, was the equivalent of €770. Only beneath the spandrel panels did the architect decide to use glued laminated timber beams with a depth of 490mm instead of 342mm. This was to suit the panes of insulating glass, which were only available at a reasonable price in a standard height of 250cm. About half the insulating glass units have a polyester fleece fitted in the cavity between the panes. One reason

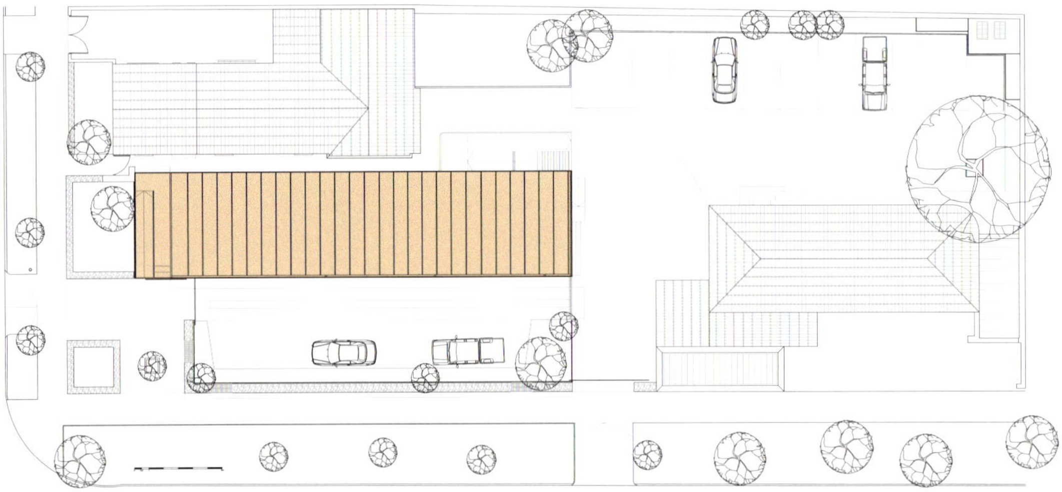

BIP Computer, site layout; top left and bottom right are the existing buildings that frame the new building at centre left. The total area of the plot is 1654m², of which 150m² are occupied by the new building.

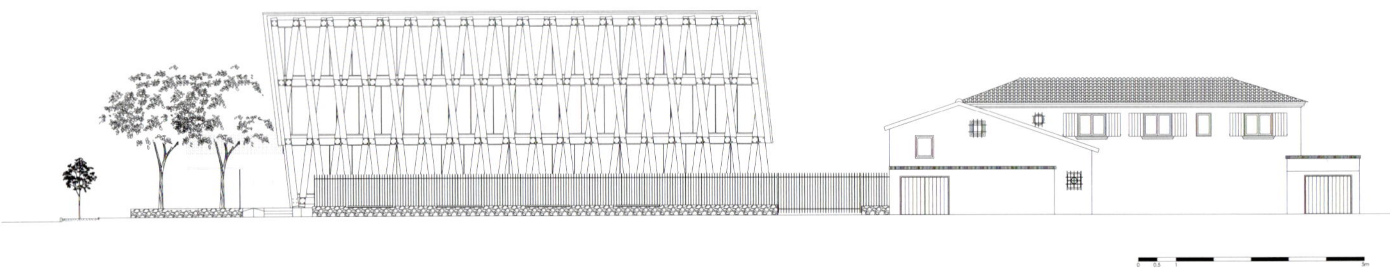

West elevation of new building and the existing building that was integrated to function as a repair and service centre.

BIP Computer, entrance area and junction with existing building. The gable end façade is clad with conventional formwork panels as used for concrete construction.

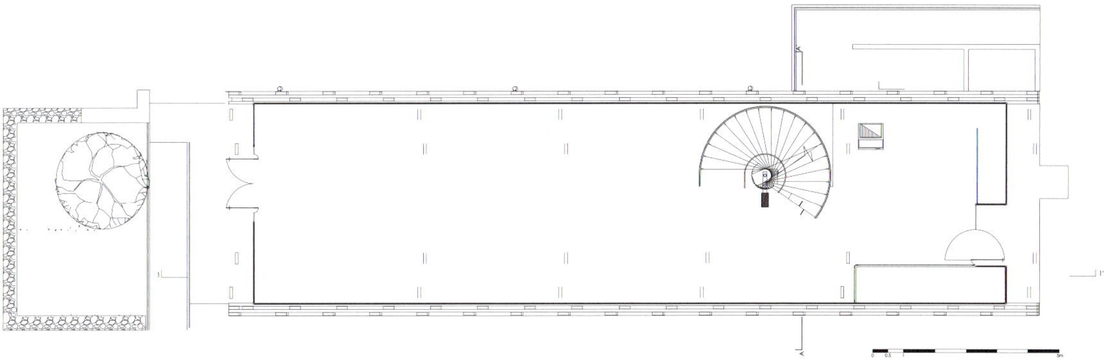

Plan of ground floor.

for this is glare protection in the areas exposed to direct solar radiation and privacy where the façade is very close to the existing buildings. Another is that, according to the architect, this measure also improves the thermal insulation properties of the glass units.

The species of wood used for the glued laminated timber elements is Monterey pine (*Pinus radiata*), which is very common in Chile and is obtained from cultivated forests in the south of the country. Urea-melamine-formaldehyde (UMF) resin was used for gluing the individual laminations in order to achieve high resistance to moisture. Painting with a "breathable" protective lacquer gave the wood its dark brown colour. This lacquer also contains a pesticide to protect against termites, which are widespread in Chile.

The basic structural idea of the collar-type diagonal column pairs was to achieve stability merely from the frame effect of the rhombus-like lattice of the façade. Every connection between horizontal beam and diagonal column has six bolts (eight for the internal spandrel panel beam) with a shank diameter of ¾ inch (19mm) such that the joint is rigid and ensures adequate stability in the longitudinal direction of the building. In order to be able to transfer the high individual forces in the bolts at such a direct connection and reduce the slip of the bolts in the holes drilled through the timber components, 6mm thick galvanised steel plates were inserted between the timber elements.

All the beam and column layers are separated from each other, i.e., beams and columns act as continuous members that are joined like collars where they intersect. Besides the structural advantages of such a design, this form of construction also eases dismantling and reuse of the individual components. It also helped to keep down the cost of the basic structure, which worked out to be only €620/m^2.

As only the longitudinal façades consist of the rhomboid lattice of diagonal column members, an independent bracing system had to be devised for the gable ends of the building. Alberto Mozó solved this problem by using the same diagonal members that make up the façade but placed these in the interior in the transverse direction. These are merely bracing, not loadbearing, members.

The internal bracing columns follow the diagonal column axes of the longitudinal façade. The upshot of

The façade construction during assembly on site. The 6mm thick steel plates at the junctions with the diagonal columns are readily visible.

Assembling the timber façade structure on the concrete ground floor slab.

Connection between internal columns and floor beam. Visible here are the crossing internal columns, which are connected directly to every ninth (in the outer bay every twelfth) floor beam.

Connection between stair and RC column. Like the rest of the structure, the spiral stair linking the three floors is made from glued laminated timber beams with a 342 × 90mm cross-section. The spiral stair is designed as a cantilevering construction with the treads fixed in the centre.

this is that they are inclined at 6.7° to the vertical and this is reflected in the gable end façades, lending the building a dynamic character.

Helped by extensive prefabrication, it was possible to complete the building in just four months. Once the concrete basement was finished, the fully prefabricated glued laminated timber members were assembled horizontally on the concrete ground floor to form the façades. Each one of the 10m long glued laminated timber members weighs 80kg and could therefore be easily moved and installed by the erection crew without any lifting gear. After the timber structures for both longitudinal façades had been fully assembled, these approx. 12t units were lifted into position by a mobile crane and secured with steel guy ropes. The floor beams were fitted into beam hangers already attached to the inner faces of the façades before the internal columns necessary for stability were erected. After this, the gable end façades were closed off above ground floor level with timber formwork panels as exterior cladding. Counter battens, waterproof sheeting, thermal insulation and, finally, plywood panels (= internal lining) followed.

The roof uses the same floor beams, but in the size 242 × 90mm, plus a secondary layer of 90 × 90mm timber beams. The roof finishes consist of sheathing in the form of 18mm thick plywood followed by 100mm thermal insulation, 15mm thick OSB and, finally, a roof covering of 6mm thick galvanised sheet steel.

The floor beams every 50cm carry floor panels in the form of 50 × 50cm conventional reinforced concrete flags that are normally used outdoors (so-called *Pastelones de Concreto*). The great advantage of these factory-produced and, above all, very low-priced products is their high thermal mass, which reduces the rapid build-up of heat in the building so typical of lightweight timber structures during the city's hot summers. Furthermore, the high density and mass of this product improves the acoustic properties of the suspended floor.

This commercial property in Santiago de Chile designed by Alberto Mozó is an excellent example of the options that modern timber construction offers: an

Floor opening, 1st floor. Like the façade columns, the internal columns continue through all three storeys, which means openings are required in the upper floors. In order to provide a satisfactory architectural solution to this problem, the openings are turned into an architectural "feature", i.e., they are larger than they need to be and fitted with steel grids for safety. An undeniable disadvantage of this detail is that it is not possible to separate the three storeys in terms of acoustics or ventilation. Like the beams to the upper floors, the internal columns are made from standard 342 × 90mm cross-sections.

BIP Computer, view from Avenida Suecia. Polyester fleece placed between the panes of glass turns the double-glazed units into translucent panels that improve privacy and reduce glare.

extremely high degree of prefabrication with all the benefits that brings in terms of cost- and time-savings, the opportunity to achieve simple and non-wasteful dismantling and rebuilding of the structure, the low weight of the timber elements used, which minimises the need for lifting gear, and last but not least, the ecological advantages of a form of construction in which CO_2 is permanently sequestered on the basis of a renewable raw material.

Constructing the floor with factory-produced concrete flags normally used in horticulture.

Lifting the longitudinal façades into position, prior to attaching the floor beams and erecting the internal columns for stability.

Clubhouse, Haesley Nine Bridges Golf Course

Yeoju, South Korea

Location	Yeoju, South Korea
Design period	2006–2009; completed: March 2010
Architects	Shigeru Ban, Tokyo, Japan/Kevin S. Yoon, KACI International, Seoul
Structural engineers	SJB. Kempter. Fitze, Eschenbach
Structural consultant	Création Holz, Hermann Blumer, Herisau
Geometry analysis	designtoproduction, Fabian Scheurer, Zurich
Roof surface design	iCapp GmbH, Zurich
Fabrication drawings & production	Blumer Lehmann AG, Gossau
Timber product used	glued laminated timber (spruce)

Simone Jeska

The restaurant, conference rooms, a spa area, small apartments, offices and VIP lounge are all housed in the new clubhouse.

The building of a new golf course in Yeoju included the erection of a new clubhouse with a very unusual, distinctive roof structure. The new three-storey structure is roofed over by a lattice of glued laminated timber beams in single and double curvature. Twenty-one tree-like columns positioned on a square, 9×9m grid support the 76×36m roof. Measuring 60cm in diameter at the base, the columns gradually widen to a diameter of 1.50m at a height of 9.60m where the individual members of the columns spread out like the branches of a tree to merge into the timber latticework of the roof[1]. At the intersections – in the middle of each 9×9m bay – at a height of 13.60m, the lattice structure of the roof forms ridge lines, and cantilevers out 4.50m to the edges of the roof to form a canopy around the whole building.

The free-form roof geometry was developed as a parametric model using the defined positions of the edge of the roof, the columns and the ridge lines plus the tangential angle at the column/roof transition[2].

Projecting a planar, regular grid onto the curved roof geometry enabled the axes of the beams to be defined. The grid consists of three beam axes at a spacing of about 90cm arranged in such a way that pairs of axes intersect to produce a network of hexagons and triangles. Extensive structural calculations and analyses – new software was required to establish the load-carrying capacity of the 20 000 nodes and 30 000 members for 30 loading cases – were needed to generate the beam geometries along the axes. Superimposing the mesh-like structure on the double-curvature roof surface led to beams with complex geometries, whose curves and twists result in a multitude of individual forms with different radii and changes of direction.

The detailing of the nodes was also affected by the complexity of the design of the double-curvature structure. In a further development of traditional scarf and halving joints, used for the first time in an engineered timber structure, the nodes represent a new departure

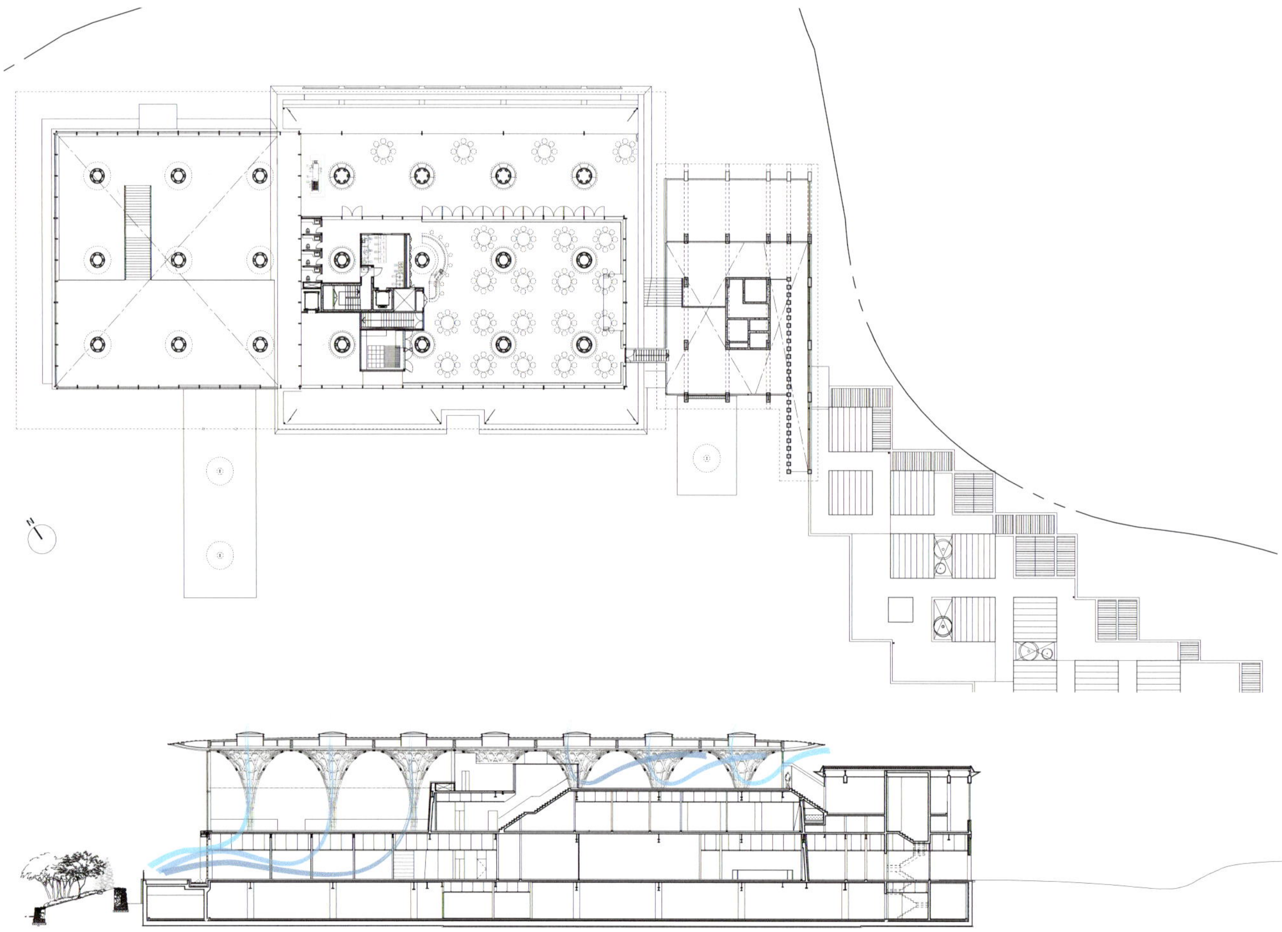

Plan and section.

1 Owing to the long buckling length of almost 14m, the individual members of the fixed-base columns are glued together to form one timber column. Additional lighting and ventilation for the interior is guaranteed by a dome-type, 3m diameter rooflight above each column.

2 Whereas the roof structure to the Centre Pompidou in Metz, the predecessor to this roof, made use of a parametric model developed on the basis of the given geometry, this time the company designtoproduction used the boundary conditions to generate the roof geometry to the clubhouse directly as a parametric geometry. Producing the parameterised data would have been too much for the design software generally in use in the construction industry and so iCapp helped to develop the digital model.

At the top of the column a central tension ring prevents the multi-member columns from spreading apart.

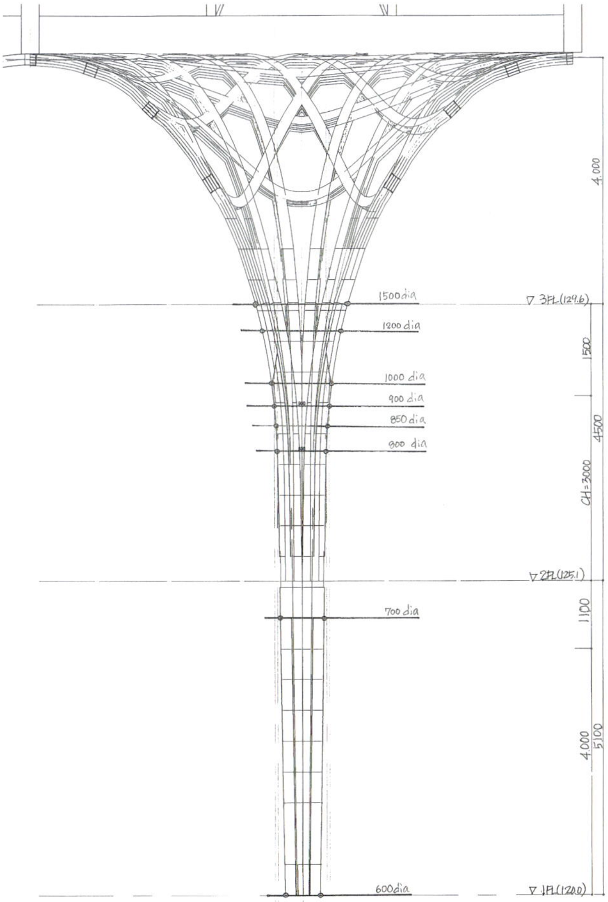

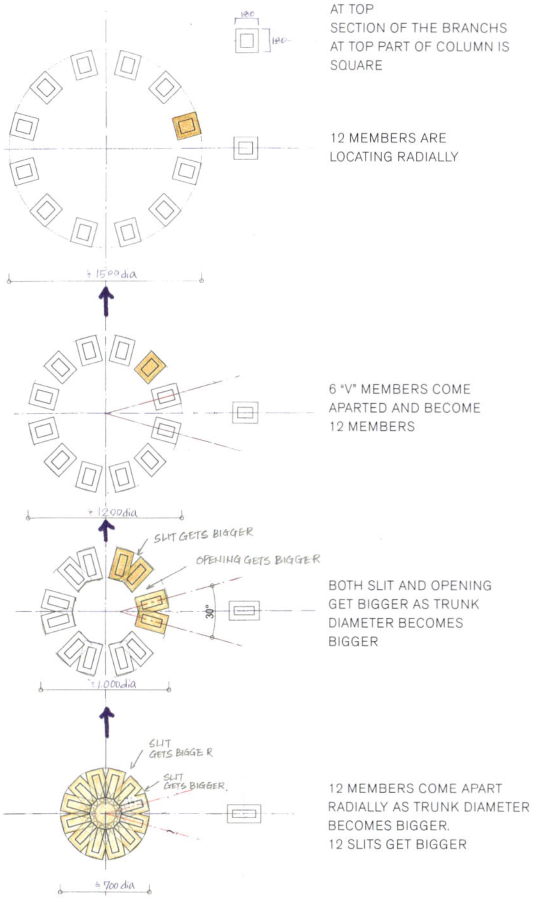

Each hollow, circular column is made up of 12 glued laminated timber members measuring 136 × 200mm and arranged in a circle. These are joined together by screw-pressure gluing.

in two senses. The concept of beams penetrating each other at the same level, which can be attributed to the high fire protection requirements, led to a return to old timber building traditions; accordingly, the longitudinal connections between the beams are in the form of scarf joints and the connections at the intersections make use of halving joints. As beams in single and double curvature meet at every node, the mating surfaces of the joints are hyperbolic-paraboloid surfaces with a multitude of different cross-sections. To increase the bending stiffness of the structure[3] and simplify erection, the beams were not provided with simple, matching notches, but instead were resolved into pairs of beams with each one having half a joint.

Whereas one pair of beams is fitted together back to back, the second pair of beams, with matching notches, crosses above and below; if this method is applied to a three-axis system, the result is three symmetrical beam pairs that penetrate each other in the same plane. To simplify fabrication, the middle pair of beams was assembled to form one beam so that the lattice could be produced and erected in five beam layers.

When specifying the segmentation of the beams, the costs of the initial timber blanks and the hours for the machining work were weighed against each other,

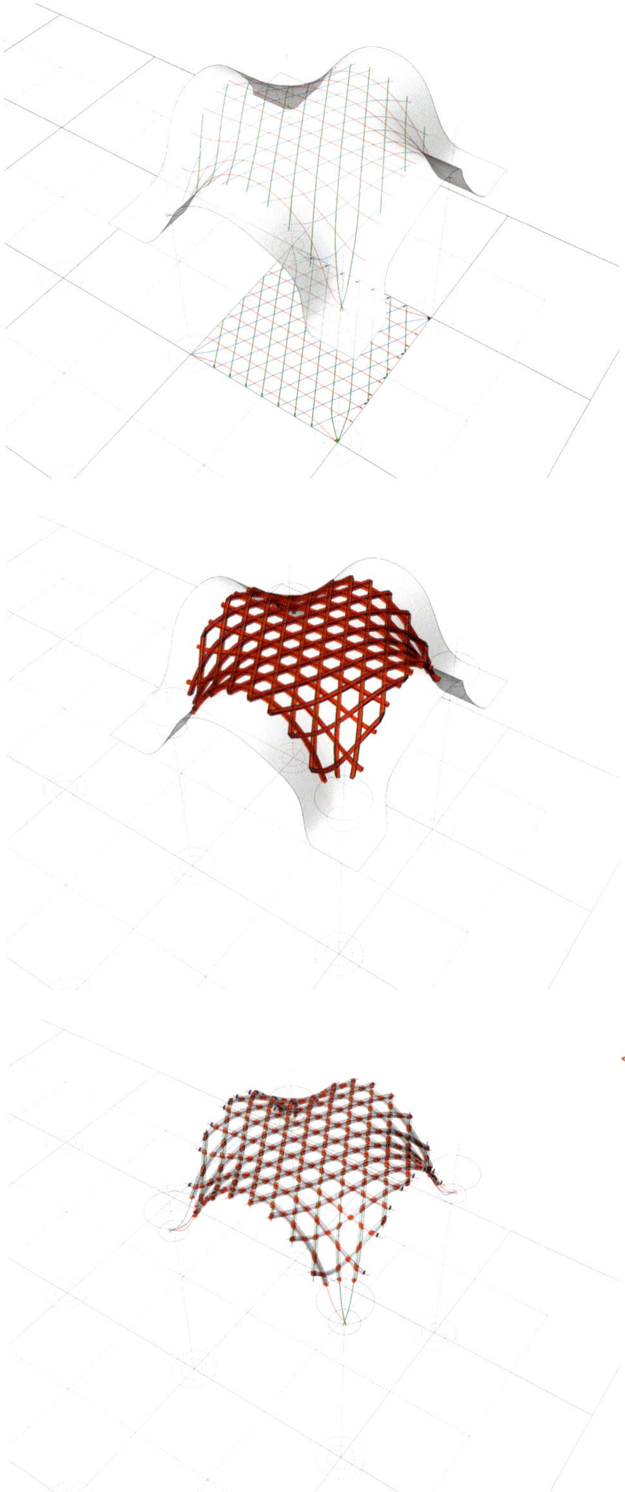

Using the reference surface of the roof geometry developed with parameters, it was possible to develop, specify and illustrate the axes, beam geometries and nodes with their three-dimensional curvature. The CAD software available at the time would not have been able to process such complex geometries.

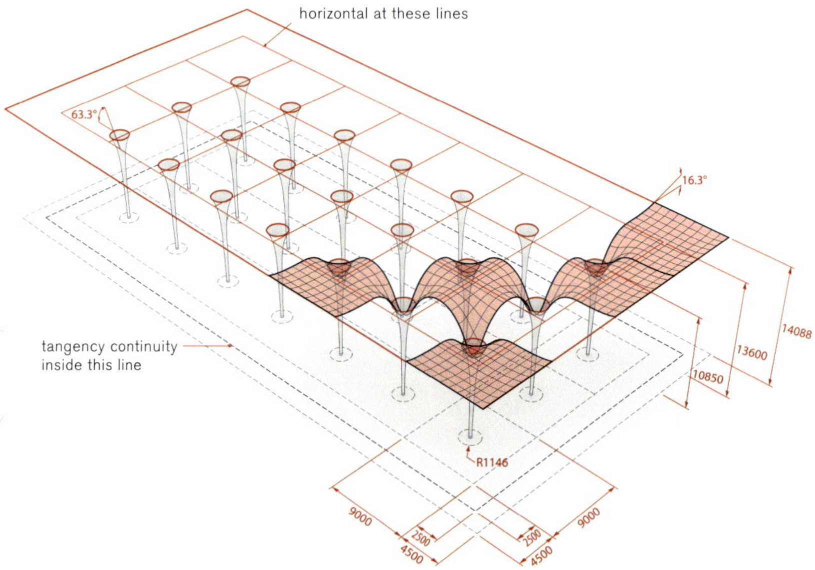

The parametric model of the roof geometry was developed with the help of defined data. Owing to the regularity of the geometry, the roof is assembled from five types of element.

the max. 10° grain cutting angle – relevant structurally – checked and the maximum component length of 11m taken into account[4]. Once the specification was established, the drawings of the individual pieces with their double-curvature mating surfaces at the halving joints and splayed cuts plus the drilling and notches required for the timber/steel connections as well as the material procurement lists could be generated automatically by the program[5].

The parameterisation, and the automation of the associated design work, enabled changes to the geometry and their effects to be ascertained at any time and taken into account, even while work was in progress. A total of 3500 beam segments, mostly in double curvature, with 476 different geometries and 15 000 geometrically complex halving joints were produced for the roof structure. Despite the regularity of the structure itself, which could be broken down into five different types of element[6], fabrication was a challenge, even at the stage of producing the timber blanks from which the members would be cut[7]. The reason for this is that the thickness of the laminations chosen for a glued laminated timber beam depends on the radius of curvature. Beams with tight radii < 1m were assembled by hand

3 Compared with a traditional halving joint, two intersecting beams each with half the notch increases the bending resistance by a factor of three in a design with resolved beam cross-sections. At the same time, resolving the beams strengthened the respective principal beam direction.

4 For transport in containers, no component could be longer than 11m.

5 Despite the complexity of this roof structure, the parameterisation of the geometry, and hence the automation of the design steps, meant that it was possible to plan, design, fabricate, transport and erect the roof in less than eight months.

6 As the underlying hexagonal grid is symmetric about a line and not a point, the result is two different edge and corner bays for the roof structure.

7 Owing to the tight timetable, the timber sections for the glulam members had to be obtained from different suppliers throughout Europe.

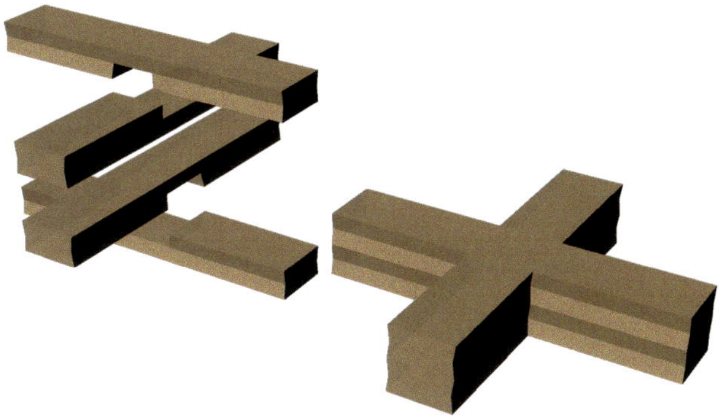

Principle of beam intersections at the same level with halving joints.

In some cases the double-curvature beams are twisted such that the ratio of the sides of the component changes continuously.

With such a complex roof geometry, every beam segment had to be designed individually to fit into the structure.

The timber loadbearing structure, which includes members with a tight curvature, is reminiscent of “woven” basket forms.

from, in some instances, 5mm thick laminations – a method customarily used when building stairs, but requiring approval by the building authorities for loadbearing components.

In order to be able to produce the double-curvature, geometrically complex beams, the five-axis CNC machining centre had to be upgraded and new software written. The lack of data interfaces between CAD and CAM called for considerable input to prepare the geometrical data for the machinery; programmers and machine operators were working around the clock in three shifts for three months so that the project could be completed in six months[8]. Another problem was the permissible tolerances, in the 0.1mm range, which came about because of the number of connections and the design employing traditional wood joints. Double-curvature beam geometries that require machining on all sides must either be rechucked once or produced from a timber section that is already curved. Therefore, the exact positioning and fixing of the curved components in the machining centre became another aspect crucial to the success of the project.

Finally, the individual components were numbered so that they could be readily identified and allocated to their place in the total structural system[9], and then sent to South Korea in 26 containers. The beam segments were assembled to form roof elements with an area of 81m² in a tent on site in which the temperature and

8 In addition, a new production building was built especially for this project.

9 In a structure assembled from thousands of differently shaped separate parts, unique labelling is essential for correct assembly. However, recording every individual component without incurring excessive costs is impossible with conventional methods, which meant that even this step had to be integrated into the parameterised production data at an early stage.

The curved beam segments are notched for the nodes. Owing to the regularity of the loadbearing structure, some beam geometries have a repetition factor of 12.

The predetermined erection sequence and the halving joints, which leave no leeway in the jointing process, meant that the five-layer beam segments aligned themselves when assembling the roof elements.

humidity were maintained at a constant level suitable for gluing. The use of digitally designed, CNC-fabricated templates[10] enabled the double-curvature beam segments to be exactly positioned in three dimensions. Using the structural location drawings, the beams were assembled layer by layer like a puzzle and joined by means of screw-pressure gluing. Verification of the load-carrying capacity and instructions for the gluing process were provided by tests carried out at the Higher Technical School of Wood, Biel, Switzerland[11].

Pre-assembly of the roof elements resulted in only 32 elements having to be joined together when erecting the roof structure.

The free-form roof to the clubhouse, which to a large extent is made up of beams in single and double curvature, sounded out the boundaries to the design and production processes and established traditional wood joints in engineered timber construction for the first time. Digitalisation plus manual skills added up to a congenial symbiosis in this project.

10 As each roof element was produced several times, it was worth carrying out this elaborate prefabrication method with the help of digitally produced templates.

11 The tests were carried out mainly to check the load-carrying capacity of the scarf joints.

The roof structure during erection.

The challenge during erection was the precise positioning of the large roof elements with a crane because 24 steel plates let into the wood at different angles and positions all had to be inserted at the same time.

At night the roof structure is reflected in the man-made pond.

Austria Center Vienna – "The Wave"

Vienna, Austria

Location	Bruno-Kreisky-Platz 1, 1220 Vienna, Austria
Design period	2005–2007; completed: June 2007
Architect	DI Christian Knechtl
Structural engineers	RWT plus ZT GmbH
Fabrication drawings & production	Buchacher Holzleimbau (now Buchacher Holzbausysteme GmbH)
Main timber products used	glued laminated timber (larch), 75mm three-ply cross-laminated timber boards (cladding)

Khaled Saleh Pascha

The new ACV entrance structure and the neighbouring buildings.

Austria Center Vienna (ACV), built in the 1970s, is a complex with facilities for congresses, conventions and all manner of conferences. Together with the neighbouring UNO City, the seat of the United Nations in Vienna, it forms a focal point for the urban development of the city on the north side of the Danube.

Vienna-based architect Christian Knechtl was appointed to draw up concepts for revamping and restructuring the area around the ACV within the scope of a study. Right from the first sketch, the architect proposed a pergola-type structure placed in front of the existing main entrance, which, together with the newly acquired vertical accent of the new structure, would also denote the main entrance. In the end, this proposal was realised almost without modification. The plans therefore also took functional criteria into account, e.g., the creation of a forecourt and a shelter to protect delegates from the weather as they arrive.

"The Wave", as it is commonly called, is a structure made completely from interlacing glued laminated timber members. On one side it is supported on the second-floor terrace of the existing building and on the other rests on 10 post bases fixed to the reinforced concrete roof of the underground car park. The dimensions are phenomenal: 32m long, 26m deep and a total height of 17m – dimensions that certainly would not be unusual for a medium-sized five-storey office building. The diamond-shaped loadbearing lattice is constructed from 70–170cm deep and 20cm wide glued laminated timber members. At the two outer edges, the diamonds are "straightened" and thus define the limits of the structure. The glued laminated timber members are

"The Wave", the new structure denoting the main entrance to the conference facilities, also forms a covered forecourt in front of the existing building.

50m long overall, but they had to be divided into two parts for transport and erection. A total of 260m³ of grade BS14 glued laminated timber was processed. Some 6t of steel were also required, most of which is in the form of connectors between the timber elements, but there are also several steel struts and ties required for stability. Additional safeguards were needed to resist the enormous wind forces prevailing at this location. Three-ply cross-laminated timber, 75mm thick, has been used to clad the façade and the roof and at the same time ensures the necessary stiffness. A total of 1100m² (or 85m³) of plywood was required.

Although the loadbearing structure with its curving diamonds appears to be very simple, the realisation called for technologies that, until recently, were unavailable. Fabrication of the timber structure was car-

A view of the timber structure at eye level.

A view of the entrance structure from the west, with the existing five-storey ACV building behind.

ried out by the Buchacher company from Hermagor in Austria. Buchacher's experience and manufacturing technologies resulted in all members being fully prefabricated in just six weeks and erected on site in another four – despite the size and complexity of this project. Preliminary and detailed drawing work was carried out with 3D CAD drawing tools, production with the help of appropriate CAM programs and fabrication and cutting the timber to size on CNC machining centres. All predrilling and routing operations were carried out during the element fabrication phase. Two opposite sides of each "diamond" are formed by essentially continuous glulam members (only divided in two for transport). The other two sides are short, separate filler pieces. At the junctions, fixings are in the form of lateral tension strengthening with steel dowels and special threaded rods from the WB system supplied by SFS. As shearing forces can occur due to the geometry of these complex, three-dimensional connections, i.e., forces acting in the direction of the connectors themselves as well, the threaded rods, which are fully recessed and hence invisible from outside, guarantee the necessary force transfer at the intersection. Erection on site took the form of first assembling a complete edge segment on the ground, adding the short filler pieces and then lifting it into position. A timber trestle roughly in the middle of the span helped with the rest of the erection. Each two-part, continuous member was positioned and erected on the trestle before the next set of short filler pieces was attached. Only one mobile crane was needed for the work on site. Ten steel post bases serve as supports on the roof over the underground car park,

A view of the underside of what appears to be an interwoven timber structure. Readily visible are the grooves in the three-ply CLT, which enable the material to be bent to follow the outside radius. The bracing in the form of steel cables and the steel struts at the top of the photo are required to resist the high wind loads at this location.

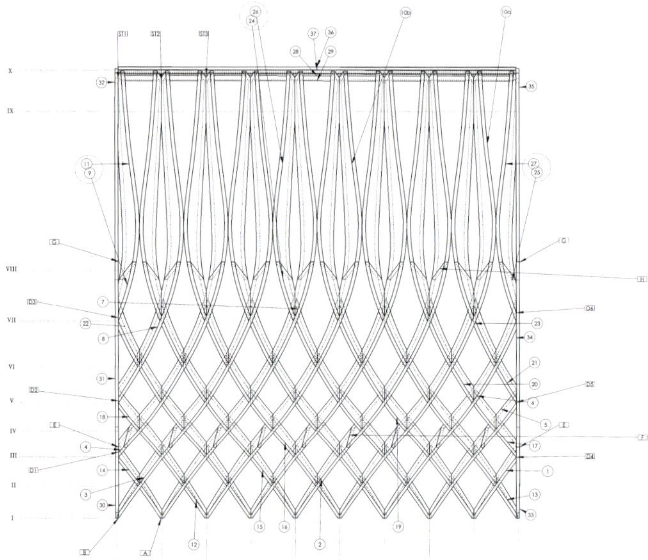

Development of the structure showing the continuous members (from left to right) and the intermediate filler pieces.

V-shaped steel columns support the other end of the entrance structure on the 2nd floor terrace of the existing building.

Owing to the relatively tight radius at the transition from façade to roof, the 75mm thick three-ply CLT panels are grooved on the inside so that they can be fitted to the curve of the structure. A polyurethane foil was sprayed onto this cladding. It is suitable for both roof and façade applications and so was ideal for this project. The advantage of this system, "Flexiskin", compared with conventional panel-type façade finishes is that it creates a totally smooth, continuous and, above all, crack-free surface. It is sprayed on and therefore also suitable for curved surfaces.

The calculated horizontal and vertical deformations are max. 10cm in both directions. As no fragile

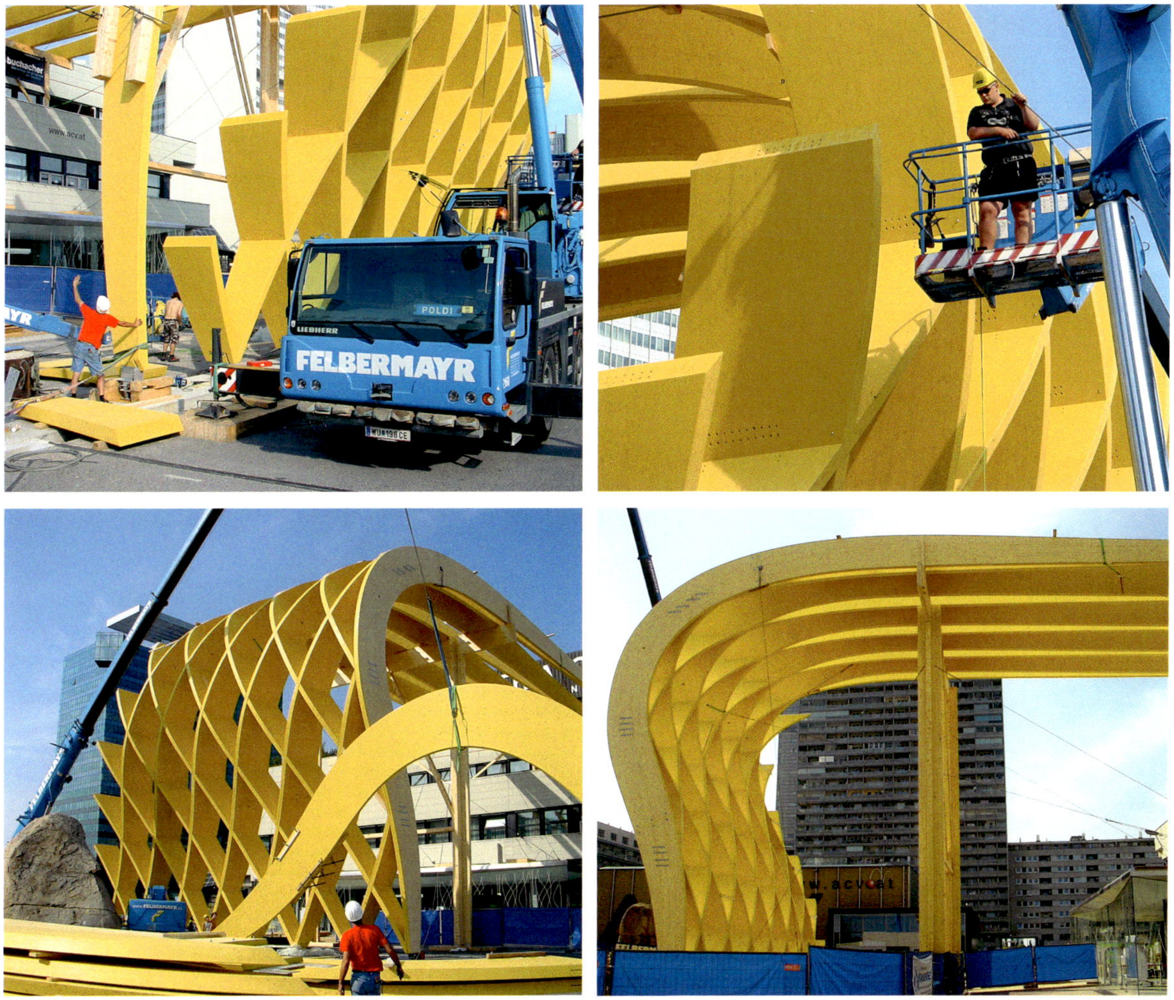

"The Wave" under construction. The members in one direction are continuous, the others are short filler segments.

materials were used in the entrance structure and it does not form an enclosing wall around an interior, which would make it more difficult to accommodate tolerances, such dimensional changes caused by expansion do not have any negative effects on the function and serviceability of the structure.

There are no services on this structure either; at night it is illuminated indirectly – no need for any light fittings on the structure itself.

In many ways the new entrance to the ACV is a remarkable example of the possibilities of modern timber construction. On the one hand, the observer is impressed by its dimensions and the fact that, despite its huge size, it seems to be much smaller and almost intimate from close-up. On the other, the elegance and simplicity of the solution are formidable. Size and elegance are attributes that we do not always associate with timber used as a building material. Curving beams 50m long with a high-quality surface finish were for a long time the province of reinforced concrete, or rather its formwork. We need only to think of the Palazzetto dello Sport by Pier Luigi Nervi, built in 1957. The precast, curving concrete columns of that structure exude an elegance and simplicity similar to the ACV project, with comparable dimensions. But the big difference in the ACV project is the material; timber lends the structure a certain scale. Although the dimensions of the primary structure are also gigantic, the grain, structure, irregularities, nuances of colour and varying surface textures of the material are references to a level of experience and scale familiar to us. In particular, seen in the context of the other buildings forming the ACV's immediate surroundings, this difference is very noticeable: the interwoven timber lattice with its sensuous, organic language wants to be "grasped" and "experienced", something that cannot be said for the neighbouring large, abstract and forbidding steel-and-glass structures.

A view of the new structure from the vestibule of the existing building.

Footbridge in Kollmann

South Tyrol, Italy

Location	Kollmann, South Tyrol, Italy
Construction	2008; completed: 2008
Architect/ structural engineer	Thomas Schrentewein, Lignaconsult, Bozen
Project team	Sebastian Vigl, Manuel Schieder, Alessandro Tombolato
Contractor	Zimmerei Robert Mauroner, Villanders
Main timber products used	cross-laminated timber (spruce) for main beams, glued laminated timber (spruce) for cantilevering deck beams

Khaled Saleh Pascha

View of north side.

This little bridge was one of the projects in the redesign of the centre of Kollmann, a village in South Tyrol. The old timber walkway crossing the Ganderbach stream has been replaced by a footbridge that can also be used by emergency vehicles if need be.

The special, double-curvature, asymmetric form of the bridge, the eccentric position of the solid main beam and the ensuing structural peculiarities led to the choice of cross-laminated timber (CLT) as the construction material – a material that had previously been used very little for bridges at the time.

The asymmetry of the bridge is not only obvious from the eccentric position of the main beam and the curving plan form; it is also evident in the width between the parapets, which varies from 2.34 to 3.06m.

The curved shape of the bridge and the position of the main beam to one side, with its cantilevering secondary beams carrying the actual deck itself, applies considerable torsion forces to the section, i.e., the timber is loaded not only in the longitudinal direction of the beam, but also in the transverse direction. CLT is especially suitable for such a task because its cross-banded lay-up means it is significantly more isotropic in its behaviour than glued laminated timber, where all the laminations are laid up parallel with the grain. The cross-section varies along the beam axis to match the internal forces.

Each of the 11 layers making up the bridge beam has a radial cutout with a different, increasing radius in its underside such that the original trapezoidal cross-section at the supports gradually changes to a triangular cross-section at mid-span, giving the bridge its characteristic organic form when viewed from the sides and below. The minimal triangular cross-section at mid-span is designed for the bending moment and at the same time reduces the torsion moment at the supports for the self-weight loading case. According to the designers, the basic idea for the special, double-curvature form of the main beam was to counteract, or at least optimise, the torsion forces due to imposed loads by varying the self-weight of the solid beam. This worked only to a certain extent. What turned out to be very advantageous, however, was the resulting trapezoidal beam cross-section at the supports, which simplified the transfer of the torsion and shear forces to the abutments. The supports are in the form of forks

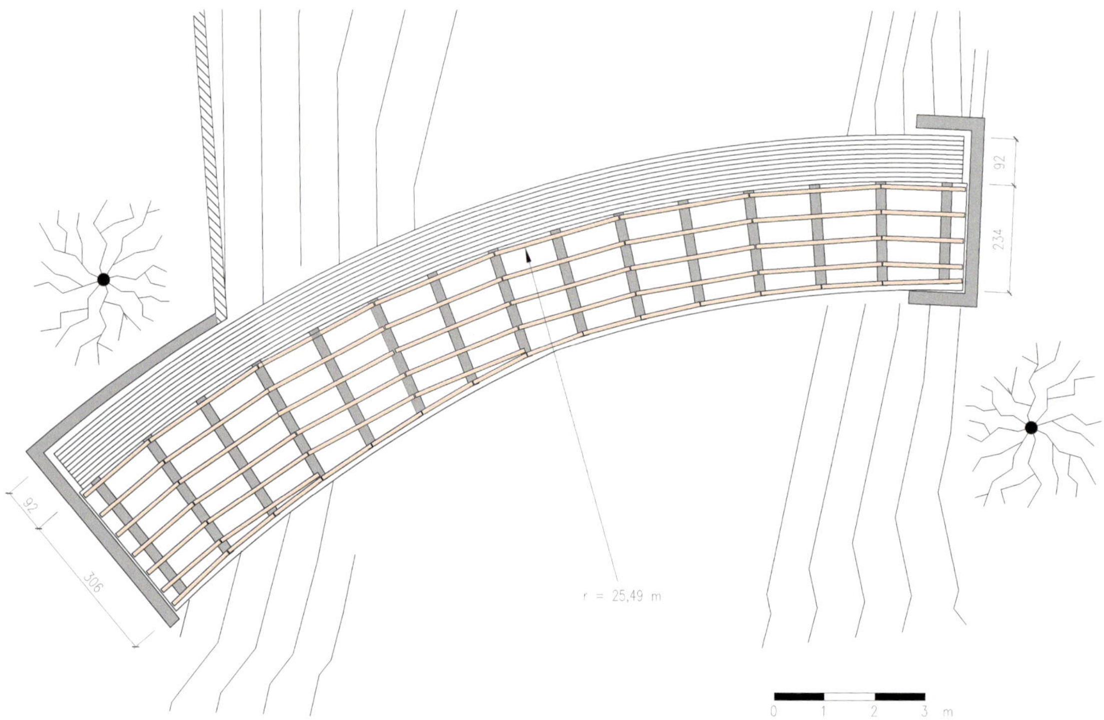

Plan of footbridge in Kollmann.

that achieve fixity in the longitudinal direction. Cast-in M30 bolts connect the steel plates of the beam hangers on the timber bridge to the supports.

Eleven layers of three-ply GL 24h CLT elements, each 84mm thick, were glued together to create the main bridge beam. The designation "GL 24h" stands for glued laminated timber with a permissible bending stress of 24N/mm² over the entire depth of the cross-section. The maximum length of the CLT elements available was 16m and so every one of the 11 layers of the 19.35m long bridge beam had to be assembled from two pieces. The desired curvature (min. 25.20m radius) was then created with the help of a support, a sort of centering. The layers were joined together one by one by means of screw-pressure gluing, in each case maintaining an overlap of 4m. Using a few screws as connectors between the successive layers, it was possible to fix the curve of each individual layer before applying the adhesive over the entire area. The fear that the curved member would try to return to its original shape after removing the support proved to be unfounded – the beam retained the form defined by the centering after assembly.

Section at l/2 (mid-span)

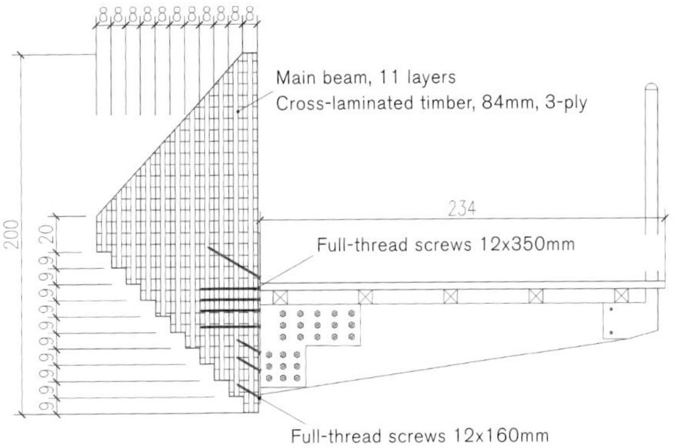

Section at l/4

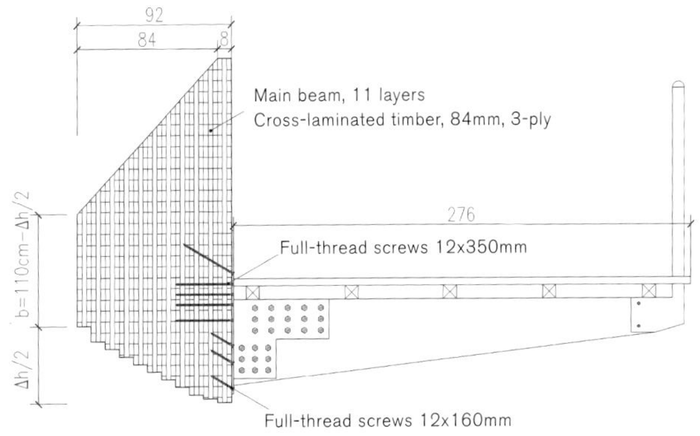

Section at l = 0 (support)

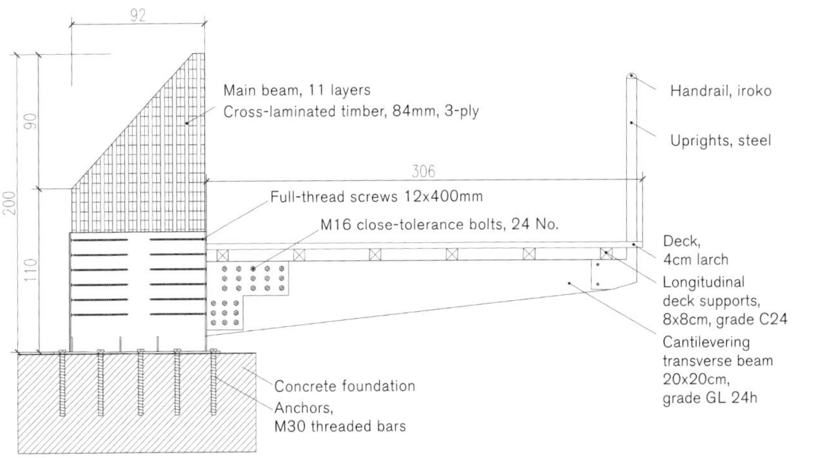

Sections through the bridge.

It took two weeks to fabricate the main beam. The work was carried out entirely in the timber contractor's works.

The fully prefabricated solid main beam, weighing approximately 15t, was transported to site on a low-loader, using the centering as a temporary support, and then lifted into position on the prepared foundations with a mobile crane. The transverse beams supporting the deck at a spacing of 1.32m were then attached. As the main beam is positioned asymmetrically, on one side of the bridge, each cantilevering transverse beam tapers to match the bending moment (deeper at main beam end). The transverse beams are connected to the main beam with full-thread 12mm diameter bolts. Once the transverse beams were in place, 8 × 8cm longitudinal beams could be attached to carry the actual deck itself, which consists of 4cm thick larch planks.

For the structural calculations, the main beam was divided into 20 equal parts. Two members therefore have the same cross-section parameters. Determining these effective cross-section parameters was very involved, especially as the moments of inertia had to be converted. Only at the triangular cross-section at mid-span was this unnecessary. One special feature of CLT must be considered in such work: the cross-plies are not allowed to be included when checking bending and shear stresses. Consequently, the natural arrangement of the cross-plies was only advantageous when checking the torsion stresses at the supports. The loading cases critical for the design of the main beam were shear and torsion at the supports and bending at mid-span.

Despite the complexity of the curving main beam, only two main construction details had to be worked out for this project: the supports with their steel beam hangers, designed for torsion and the support reaction, and the connections between the transverse beams and the main beam, which use steel flanges designed for the fixed-end moment and the shear force. At the support, the prefabricated steel parts were combined into one.

Cladding in the form of two layers of larch shingles protects the main beam against the effects of the weather. This type of cladding can be found on many old bridges in this region. In addition, a plastic membrane protects the bridge beam against moisture. A delicate steel parapet is fitted to the outer ends of the cantilevering transverse beams. On the other, northern side of the bridge, the main beam itself functions as a parapet and safety barrier.

The footbridge in Kollmann is an excellent example of the architectural and engineering options that result from the use of modern timber materials, in this case CLT. Asymmetric solid beams, the safe transfer of bending forces occurring transversely to the axis of the

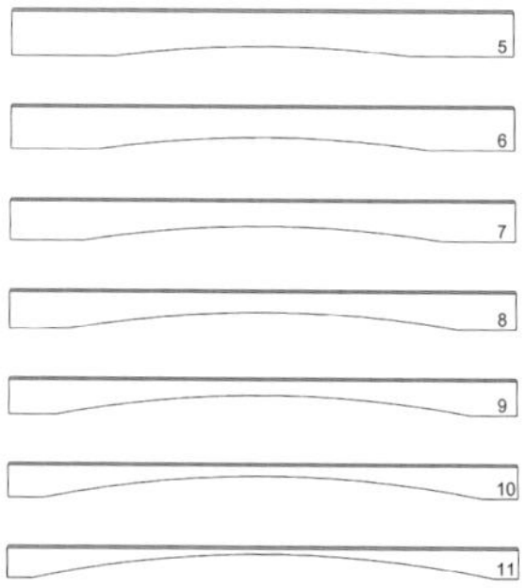

Elevations on the developments of the individual layers: number 1 is on the south side of the bridge, where the cantilevering secondary beams carrying the deck are attached, number 11 on the north side.

Fabrication and erection of main beam: Joining together the individual layers over centering used to define the desired radius. Fabrication of the glued main beam in the factory. Transport and erection of the beam.

member, limiting shrinkage by using very large, prefabricated timber single-span beams and, on the whole, a very sturdy structure are the attributes of modern CLT technology. These constructional qualities lead to a new architectural freedom in which the traditional image of the timber structure, as a highly directional form of construction defined by grid-lines, gradually gives way to a dynamic aesthetic that follows the bending moments and shear forces. In this sense, modern construction with solid timber is becoming similar to that of reinforced concrete, with the big difference being that timber needs no elaborate formwork. As the example of the footbridge in Kollmann demonstrates very elegantly, geometrically challenging forms can be created by combining dissimilar, two-dimensional elements and exploiting the elasticity of timber.

Cantilevering secondary beam and underside of deck. Positioning the main beam on the abutments. The main beam prior to adding the shingle cladding.

Gessental bridge

near Ronneburg, Germany

Location	Gessenbach valley near Ronneburg (BUGA park), Germany
Completed	2006
Architects/ structural engineers	Büro für Ingenieur-Architektur, Richard J. Dietrich, Traunstein
Static and dynamic design	Köppl Ingenieure, Rosenheim
Checking engineer	Josef Trabert, Geisa
Fabrication drawings & production	Schaffitzel Holzindustrie GmbH + Co., Schwäbisch-Hall
Timber product used, stress ribbon	glued laminated timber (spruce, laid up in parallel blocks)

Simone Jeska

Gessental bridge carries a long-distance cycle trail linking several towns in Thuringia across the 25m deep and up to 300m wide valley of the little River Gessenbach.

Gessental bridge was built as part of the preparations for the 2007 German National Horticultural Show (BUGA) in Gera and Ronneburg to link two parts of this major event. Based on experience gained on Essinger bridge[1], Gessental bridge was designed as a 225m long stress ribbon[2] with a homogeneous loadbearing ribbon[3] made from glued laminated timber laid up in parallel blocks. Both in elevation and on plan, the bridge has a gently undulating form. On plan, the width of the deck varies continuously between 2.90 and 3.90m, with the wider sections being over the supports, i.e., at the high points of the stress ribbon. In elevation, the wavy line of the bridge ensues due to the high points: the concrete abutments, where the ribbon is anchored, and the four intermediate piers (arranged in pairs) made from steel circular hollow sections, which act as saddle points. With "half-wavelengths" of up to 80m, the tree-type branching at the top of each pier leads to clear spans of 52.50m for the two side spans and 55m for the centre span. Between the supports, the shape of the stress ribbon adjusts itself to suit the catenary curve. The optimum catenary curve is determined by achieving a reasonable balance of function, structural efficiency and economy. Choosing a large sag reduces the tensile forces, which, on the one hand, has a positive effect on the dimensions of the abutments and the

1 The 193m long timber bridge over the Main-Danube Canal near Essing can be regarded as the predecessor of the Gessental bridge. Essinger bridge, completed in 1986, was also designed by Richard J. Dietrich and was the first bridge with stress ribbons of glued laminated timber – made possible by further developments in gluing technology. The bridge has four spans and consists of nine glulam beams (22 × 65cm), boards laid diagonally on top and a diagonal wind girder underneath.

2 Stress ribbon bridges are suspension bridges whose cables are anchored in the banks. Their origins lie in crossings that use natural fibres to bridge over deep gorges.

3 The possibility of building such a stress ribbon using glulam beams laid up in parallel blocks had not yet been approved by the building authorities at the time Essinger bridge was built.

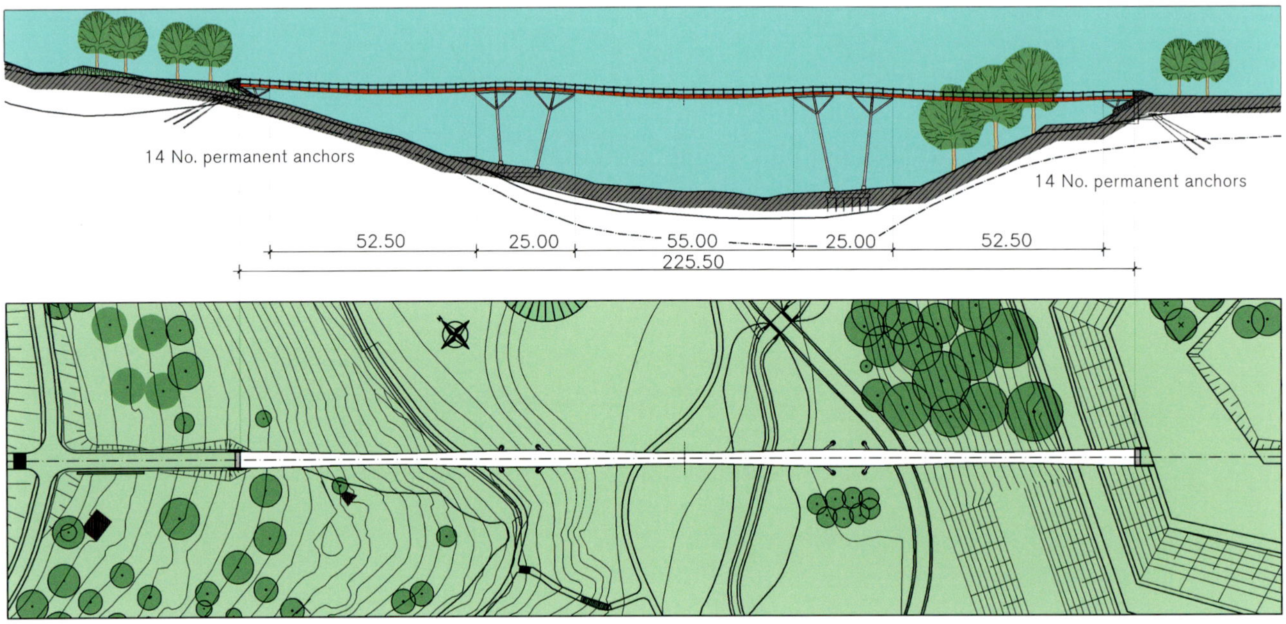

The 225m long stress ribbon bridge at the National Horticultural Show near Ronneburg, Germany – elevation and plan.

bridge cross-section but a detrimental effect on the bridge's stability. From the structural viewpoint, the optimum catenary curve for the 50cm deep stress ribbon of the Gessental bridge was achieved with a sag of 2.20m in the centre span and tensile forces of about 800t (8100kN), which must be carried by the material and the connections and resisted by the abutments. However, because such a stress ribbon would have exceeded the maximum gradient of 6% permissible for barrier-free access and a smaller sag would have led to very high forces at the abutments and hence higher costs, the timber ribbon was doubled up. Whereas the vertical alignment of the loadbearing stress ribbon corresponds to the optimum structural form, the longitudinal gradients of the deck are established by raised, 51mm thick laminated veneer lumber (LVL), Kerto-Q boards. The LVL boards are supported on four glued laminated timber beams arranged parallel with the direction of span and bolted to the stress ribbon itself.

Another positive effect of doubling up the construction, apart from optimising the catenary curve, is an improvement to the vibration and torsion behaviour of the structure. Suspension bridges are slender struc-

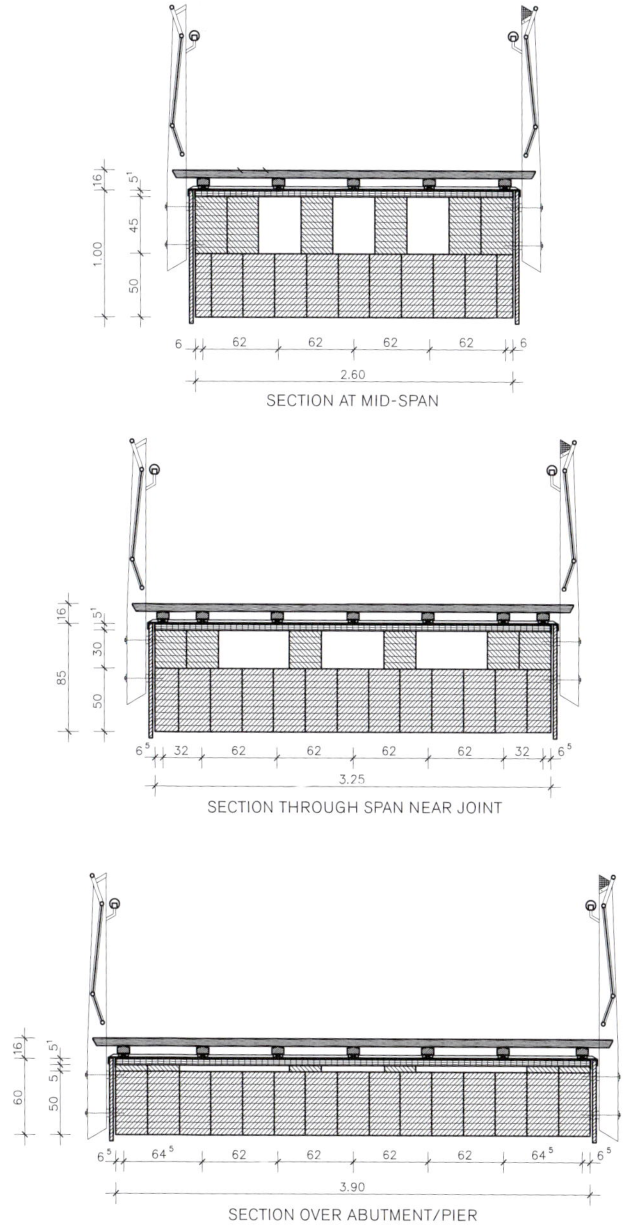

The width of the stress ribbon varies continuously between 2.90 and 3.90m.

Tapering glulam beams accommodate the gap between stress ribbon and deck, which varies between 10cm at the supports and 50cm at mid-span.

tures, which means that their vibration behaviour in response to the dynamic loads from users and wind, plus their torsional stiffness, are critical to their stability. However, the box-shaped cross-section of this bridge, brought about by the doubled-up construction, helps to counteract the torsion and the vertical vibrations in the spans. In addition, the bolted connections between the glulam beams and the stress ribbon generate friction, which damps the vertical vibrations. Narrowing the bridge at mid-span increases the torsional stiffness and the wider sections at the high points serve as viewing platforms.

To resist the horizontal loads, mainly due to wind, the steel circular hollow sections of each intermediate pier are arranged as an upturned "V" in the transverse direction and also inclined in the longitudinal direction. To transfer the tensile forces to the abutments and avoid intermediate restraints in the structure, the supports permit longitudinal movement (in this case they are designed as knuckle leaf bearings), but the stress ribbon is rigidly connected in the transverse direction. Furthermore, the elasticity of the steel circular hollow section columns, up to 23m high, also permits a certain degree of movement in the longitudinal direction.

The 225m long stress ribbon is made up of nine segments, which were prefabricated at the works. As each 30m long segment, weighing about 26t, could not be fabricated on a CNC assembly machine because of its huge size, every segment was machined manually. With a tolerance of about 1cm over the entire length of

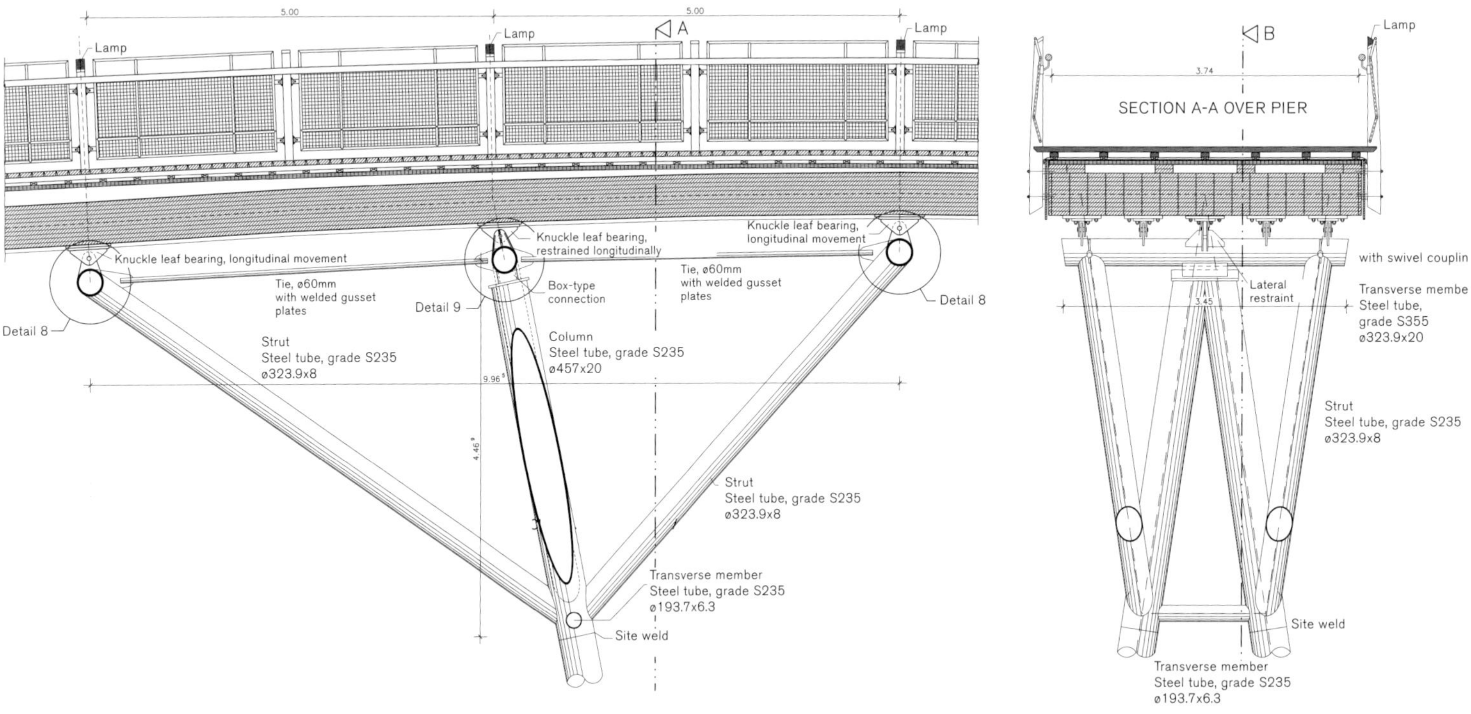

At the top of each tree-like pier there are three support points spaced 5m apart in the longitudinal direction of the bridge. The stress ribbon therefore arches over six axes at the supports, which creates a gentle, continuous rounding between the catenary curve.

the bridge, this type of production represents a serious challenge for manual operatives and designers. The beams laid up in parallel blocks are made from vertically glued, curving glulam beams of strength classes GL 28h and GL 32h with cross-sections of about 20 × 50cm, which were cut with hand-held circular-saws to match the curvature of the plan form[4] and provided with horizontal slits in the end faces for fitting the steel connecting plates.

On site, the bridge segments were connected together by plates welded to the end plates attached to the steel plates fitted in the slits. The concrete abutments are anchored in the rocky subsoil with 14 GEWI ground anchors up to 18m long.

4 Cutting the shape along the sides was carried out from above and below because the maximum depth of cut of the saws was 25cm.

Pairs of delicate steel circular hollow section piers allow the tensile forces to be transferred to the abutments.

The bridge was erected successively, starting at one abutment and supporting the deck on temporary scaffold towers. In order to establish the exact height of the scaffolds, the erection line of the stress ribbon was calculated beforehand taking into account the flexibilities and tolerances of the connections. Over the whole length of the bridge, adding together the tolerances results in a change in length of about 3cm, which would mean an additional sag of 15–20cm. The scaffolds were up to 23m high and their stability had to be checked because of the wind loads in this region. Acting as "temporary supports", the scaffolds carried the vertical loads of the bridge segments, which functioned as simple beams in bending during erection. Once erection was complete, the scaffold towers were lowered so that the stress ribbon could become tensioned and take up the calculated catenary curve.

Protection against the weather is crucial when using loadbearing timber components outdoors. Accordingly, the timber ribbon was waterproofed above the LVL board and protected by standing seam sheets laid diagonally. On top of the metal sheets, a grid of larch battens forms the actual deck, which can be easily renewed at any time. The side faces of the bridge are protected by three-ply core plywood boards (larch) in front of a ventilation cavity.

Normally, stress ribbon bridges, loaded primarily in tension, are made from reinforced concrete or steel.

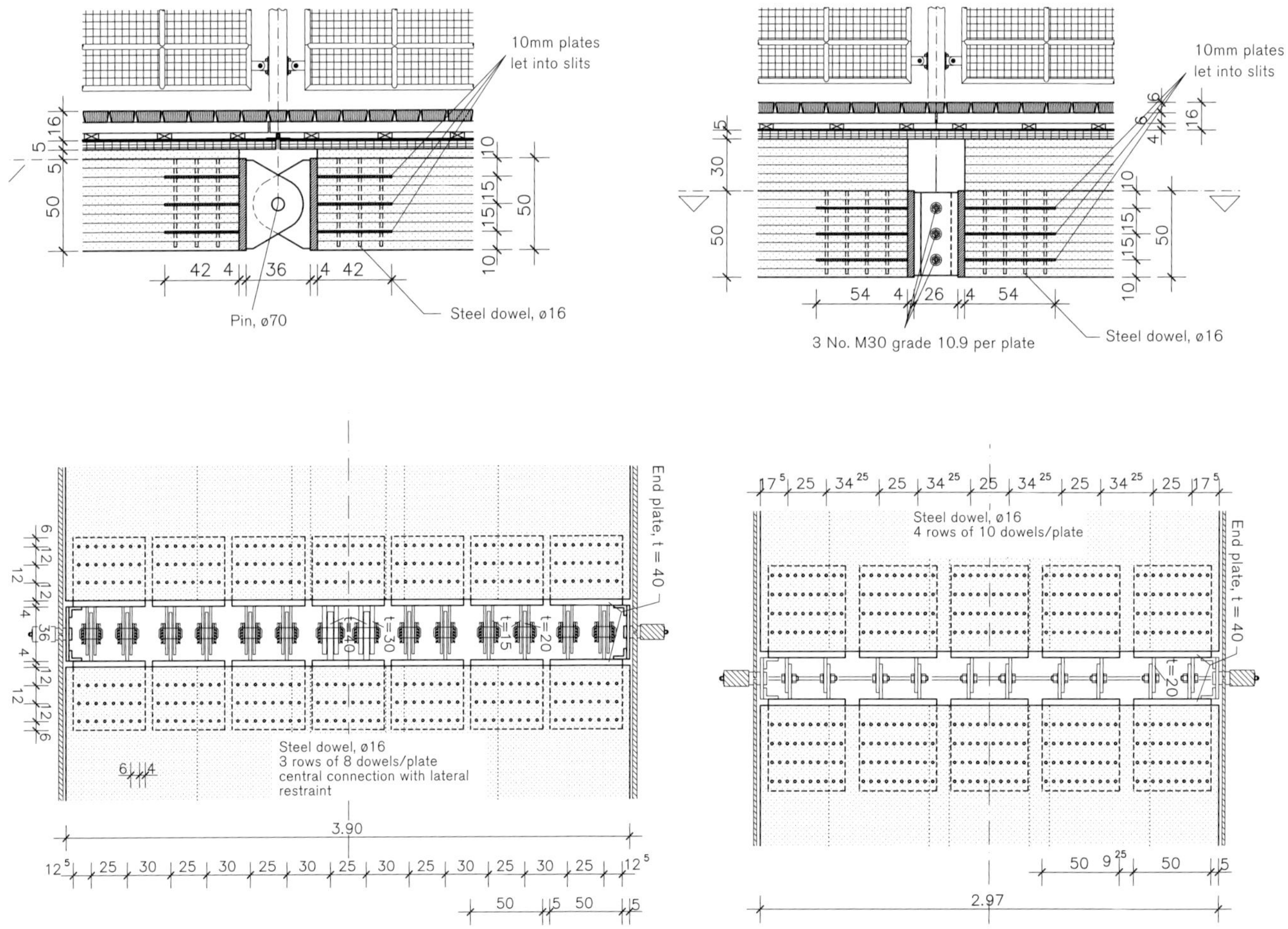

The connecting plates of each bridge segment are fixed with up to 400 steel dowels (16mm diameter) in order to transfer the high tensile force of 8100kN.

The reasons for the wave-like plan form are functional and constructional, and not merely formal.

Therefore, the decision to opt for timber seems unusual at first sight. However, a brief reflection reveals the advantages of a timber loadbearing structure over steel or reinforced concrete: the low self-weight compared with reinforced concrete reduces the tensile forces at the abutments by two-thirds[5], and owing to its inherent damping, which counteracts rhythmic vibrations and hence the uncontrolled escalation of vibrations in a suspended construction, timber is superior to the other materials[6]. Despite its slender design with a board thickness of just 50cm, the vibrations of Gessental Bridge have remained well below critical value – even under the load of the many visitors to the horticultural show.

5 Originally, the stress ribbon was to have been built in reinforced concrete. However, the construction of the abutments for this "heavyweight" structure would have led to a substantial budget overrun.

6 Steel suspension bridges are usually fitted with mechanical vibration dampers in order to prevent vibrations from escalating out of control.

The holes for the 460mm long steel dowels were drilled in the beams to an extremely high accuracy using templates and twist drills.

Pre-assembled bridge segments: dowelled connections in multiple shear with three horizontal steel plates constitute the longitudinal connections between the segments. In order to guarantee the accuracy of the connections, when installing the connecting plates, each neighbouring bridge segment was temporarily connected and positioned with the help of round steel bars.

The beams, up to 3.90m wide, were glued in special presses designed for a parallel block lay-up.

Elephant house, Zurich Zoo

Zurich, Switzerland

Location	Zürichbergstrasse 221, 8044 Zurich, Switzerland
Completed	spring 2014
Architects	Markus Schietsch Architekten GmbH, Zurich
Structural engineers	Walt + Galmarini AG, Zurich
Parametric modelling	Kaulquappe GmbH, Zurich
Timber construction	Elefantenpark Holzbau Consortium – Implenia Bau AG Holzbau, Zurich, and Strabag AG Holzbau, Lindau
Timber product used	three-ply cross-laminated timber boards

Simone Jeska

The free-form roof shell made from CLT boards covers an area of about 6000m². Employing untreated timber for the shallow undulating form helps the shell blend into the surrounding forest scenery.

The addition of an elephant compound to Zurich Zoo also involved the construction of a new, spacious building for the indoor enclosure which blends in well with the surrounding hilly landscape. The characteristic feature of the elephant house is its roof structure, which is designed as a shallow, curving timber shell.

The free-form timber shell is amorphous on plan, has a diameter of about 80m and covers an area of about 6000m². In terms of its geometry, it corresponds to a shallow, "deformed", parabolic dome with an off-centre crown about 18m above the ground and an undulating edge whose distance to the ground varies between 80cm and 10m. The high points of the edge are the locations of the entrances and exits for the visitors and the connections between the internal and external elephant enclosures.

The roof geometry is based on a shell loaded exclusively in compression, which was developed by reversing a suspended membrane model. The suspended model was simulated in the computer in order to optimise the structural form of the roof using the defined low points and the maximum building height. Groups of

Plan and section: The timber shell unites the indoor enclosure with the visitor area, the stalls and a two-storey events area under one roof. The two-storey events area with platform for spectators can be hired out for private functions.

between seven and nine 40–60cm wide and approx. 2m deep reinforced concrete columns plus a 5m high curving concrete wall support the shell at the low points of the roof edge[1]. Forces are transferred to the supports via a prestressed concrete ring beam – a member twisted in three dimensions along the edge of the timber shell.

Daylight floods into the spacious interior through 271 irregular polygonal rooflights made from three-layer, transparent ETFE foil cushions. The many openings in the roof structure (amounting to about 35% of the

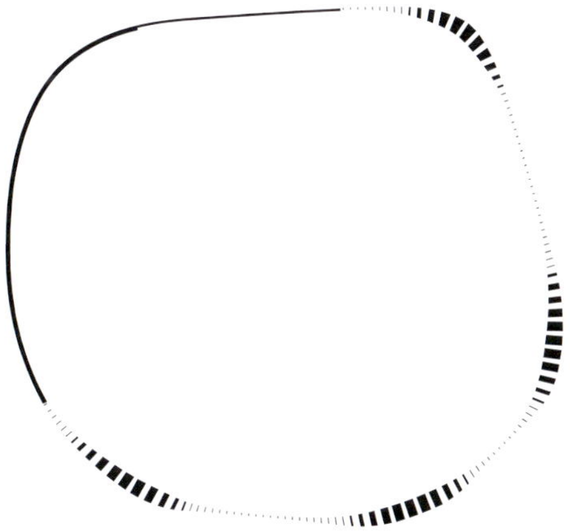

The loadbearing structure continues in the slatted form of the timber façade. Together, roof and façade coalesce into a consistent system with fluid transitions between the closed zones with concentrated loads and the open, transparent zones that allow interior and exterior to merge.

1 Horizontal thrusts in the order of magnitude of about 500t act at every support point. Rock anchors and 70cm diameter bored piles carry the loads.

The closely spaced slat-type façade posts continue the rhythm of the loadbearing structure.

surface area) resolve the shell into a network of intersecting "rays" that become more numerous as they approach the edge of the roof, in line with the flow of forces but also the lighting requirements.

Instead of using concrete, the usual choice, solid timber elements have been used to build the shell structure – the first time this has been done. However, in contrast to concrete, timber is an anisotropic material determined by its capability to carry loads parallel with the grain. So in order to guarantee a uniform load-carrying capacity in all directions, the timber shell consists of three layers of 80mm thick cross-laminated timber (CLT) panels. The layers are nailed together and each is turned through 60° with respect to the next one, corresponding to the positions of the supports.

Timber ribs bolted to the upper side of the shell give it the necessary stability. Attached in the middle of the CLT rays, the ribs join the supports together and therefore strengthen the principal loadbearing directions. Around the rooflights, the ribs strengthen the edges to the openings. The timber ribs are made up of three-layer, 8cm thick timber sections bent to suit the

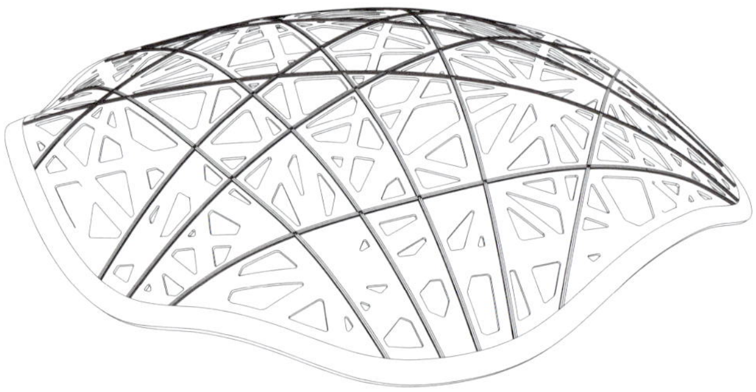

The "rays" in the principal loadbearing directions of the roof shell define the outlines of the irregularly shaped rooflights. With so many large, irregular perforations, the shell structure looks like randomly created wickerwork and awakens associations with structural forms found in nature.

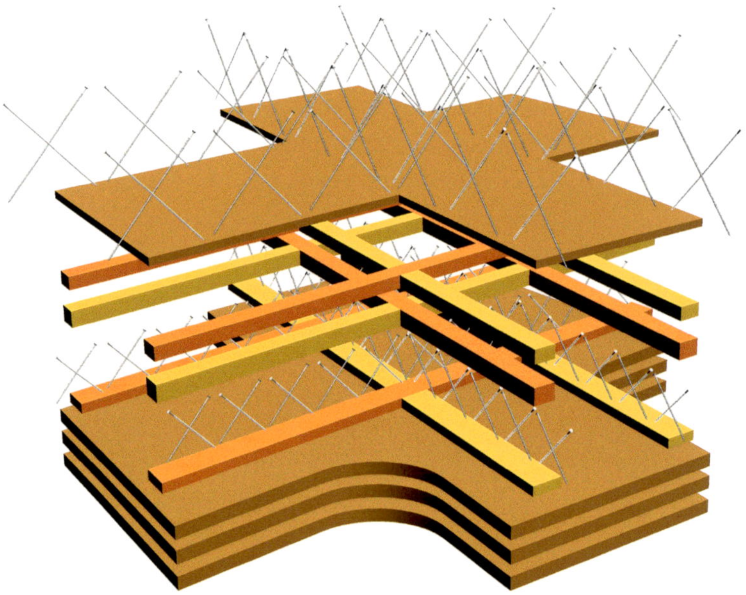

Attaching the layers of secondary ribs at the intersections was carried out according to rules that were specified in advance for the interruptions and continuations. Self-drilling screws 85cm long and installed at 45° connect the upper LVL boards with the timber ribs and the CLT elements making up the timber shell below.

curvature of the roof during erection. Where they intersect, the main ribs (24 × 32cm) are joined together with steel plates and each joint filled with grout. The secondary ribs (24 × 16cm) intersect each other at the same level owing to the alternation between interrupted and continuous layers of ribs. On top of the loadbearing shell is a 57mm thick laminated veneer lumber (LVL) board attached to the lower layers with self-drilling, angled screws[2] in such a way that the result is a honeycomb-type hollow box – a seven-part, flexible, composite cross-section with a total depth of 54cm.

As no structurally relevant performance data were available for such a form of composite construction with extremely flexible nailed/screwed connections and open joints, the structural behaviour was checked in a series of tests[3] and the values obtained entered into a finite element structural model. Based on this model, a parametric geometry model was developed which functioned as the reference drawing for the following steps in the design work. The parametric model enabled the geometries of the ribs, the ring beam and the centering to be generated automatically, and also served as the basis for refining the structural model. The three-dimensional digital model was broken down into 2.90–3.40m wide strips for each layer of CLT in order to determine the geometry of every individual piece[4], and the strips then converted into flat surfaces. In order to convert the double-curvature surfaces into flat surfaces, the designers used methods from membrane construction and modified that software to take account of the material properties of the timber elements[5].

2 The self-drilling, 13mm diameter SFS screws in a custom length of 850mm join the pieces together to create a composite construction. With their high pull-out strength, the screws can transfer shear forces. At the same time, their special coating ensures better protection against corrosion.

3 Checking the strain stiffness of the shell in compression was especially important. Six compression tests, one tension test, three shear tests and two bending tests were carried out at the Swiss Federal Laboratories for Materials Science & Technology. The results of the tests revealed a similar behaviour for loads in tension or compression and that the design, despite the great flexibility of the joints, has an adequate bending stiffness due to the multiple layers. The bending stiffness is only about 30% of that of a comparable cross-section with rigid connections.

4 The differences between the strip widths resulted from the respective roof curvature, which owing to the limitations of transport could not exceed a maximum developed width of 3.40m.

5 As surfaces in double curvature cannot be developed exactly geometrically, the calculated developments were checked for deviations in the parametric model and the conversion parameters optimised. A model of the roof shell at a scale of 1:20 was built first to verify the calculated cutting of the elements and the bending behaviour.

Translucent ETFE foil cushions close off the irregular holes in the shell.

The timber elements are joined together with 22cm long nails to form a shell structure. Using a template enables the spacing and number of nails to be maintained exactly – 100 nails/m^2. The advantage of using so many nails as connectors manifests itself in the ductile behaviour on the one hand and the robustness of the construction on the other; if one nail fails, its load is easy distributed among its neighbours.

The cut surfaces of the elements were subsequently converted into continuous, curved splines so that the geometric data, only available in the structural program as triangular meshes, would be suitable for processing on machinery. The result was that every piece had an individual cutting geometry with curved edges and irregular cutouts.

Starting at the crown of the dome, the three-layer CLT panels were bent, layer by layer, over centering made from parallel timber ribs (8 × 25cm) at a spacing of 2.57m and placed transversely to the bottommost layer of elements. The alignment and the cutouts in the top edges of the ribs[6], which match the complex geometry of the roof, were generated automatically with the help of the parametric model. Just the weight alone plus the geometrically exact cutting enabled all the pieces, up to 3.40m wide and 15m long, to be installed to match the given form. In order to create the defined free-form geometry and also guarantee adequate stability during erection, the bottom two CLT layers were laid over the entire area, without any openings for rooflights. Digitally prefabricated apertures tracing the final geometry of the full openings were initially cut in the uppermost layer of elements only. To create structural connections between the CLT elements in the longitudinal direction, the ends of the elements were routed to a wave shape, reinforced with screws crossing at an angle and the joints filled with grout. In the transverse direction, 20mm wide, open joints prevent a direct force transfer between adjoining elements and hence an unfavourable flow of forces perpendicular to the grain of the timber.

Before visitors reach the elephant compound, they pass along a path winding down a slope through Asian-

6 The ribs were 2.60–3.00m long and were supported on 880 scaffold tube posts whose levels were taken from the digital model and set out on site with tacheometers.

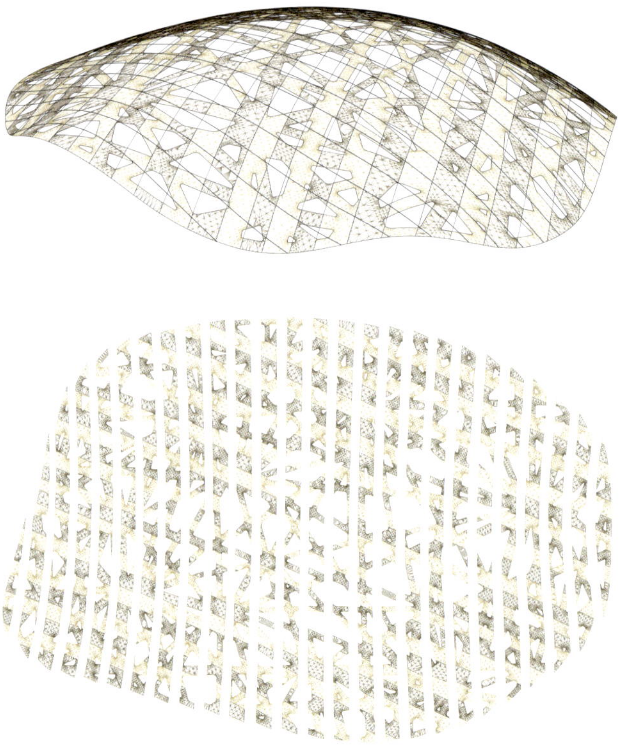

The geometry of the pieces for cutting was determined from the developed CLT element strips.

About 600 CLT panels in single curvature with widths between 3.20 and 3.40m and lengths between 10 and 15m were individually cut to size by a digitally controlled robot.

Just like the pieces in a puzzle, every one of the glulam panels is unique and only fits at one particular place in the loadbearing structure.

Elevated timber ribs define the form of the roof. With layer thicknesses of 35, 10 and 35mm, the three-ply CLT elements have a clear loadbearing direction.

A drawing showing the position of every individually fabricated element enabled each one to be fitted in the right place.

style forest and river scenery. From here they can see the free-form roof shell from above, with its honeycomb-like timber elements that form the topmost, protective layer of the roof construction. Despite its oversized dimensions, the new structure does not seem out of place. The natural material, without any finishes, the design, which takes up the undulations of the surrounding landscape, and the colours, which have been coordinated with the local environment, are responsible for the fact that the elephant house is a harmonious addition to the zoo's landscape.

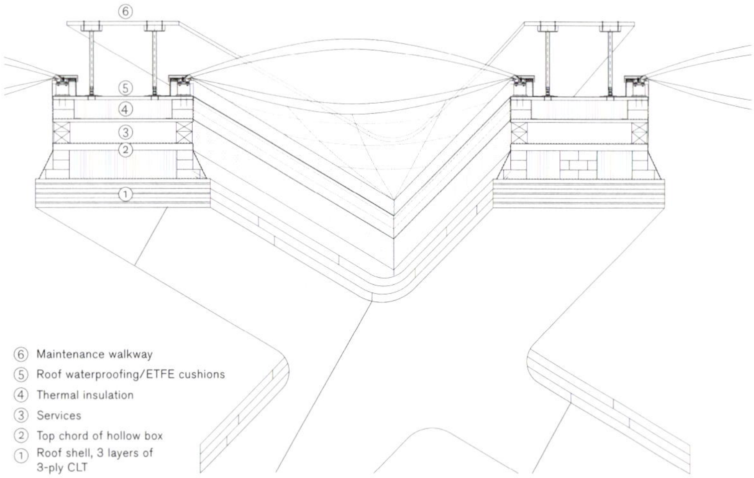

Isometric view of shell make-up: With space for services, thermal insulation and raised timber elements forming the uppermost layer to the roof, the total depth of the roof construction exceeds 1.5m. There is an additional layer of transparent plastic sheeting to protect the ETFE foil cushions against hail.

Exterior view: LVL boards 33mm thick and 60cm above the roof waterproofing on discrete supports form the topmost layer of the shell. The honeycomb-like structure of the elements is reminiscent of the shell of a tortoise.

Double sports hall

Borex-Crassier, Switzerland

Location	Rue de la Tour 55, 1263 Crassier, Switzerland
Construction	2005–2007; completed: 2007
Architects	Graeme Mann & Patricia Capua Mann, Lausanne
Structural engineers	AIC Ingenieurs-conseils SA, Lausanne
Fabrication drawings & production	Zaugg AG, Rohrbach
Timber products used, lattice trusses	hybrid design, glued laminated timber and solid structural timber sections (spruce)

Simone Jeska

Owing to the topography of the terrain, which includes a storey-high change in level, the full 9m height of the building is only apparent on the southern elevation facing the rural landscape.

The construction of a new double sports hall on the edge of the village of Borex involved rediscovering a form of timber construction that had fallen into disuse but is now responsible for the eye-catching appearance of the interior. Designed as an extension to an existing sports centre, the new building continues the lines of the existing one and is joined to it by a two-storey structure. The sports facilities are accessed from the northern side via the central building linking old and new, although the entrance at ground level leads on to a gallery above the playing area itself in the extension.

The 32 × 28m uninterrupted playing area is enclosed on three sides by 5.80m deep lattice trusses. On two sides, the almost storey-high trusses are supported on the concrete walls to the ancillary rooms. On the third side, however, the lattice truss spans over an approx. 2.50m high panoramic window providing a view over the landscape beyond.

The origins of the timber lattice truss can be found in bridges of the 19th century. It was the American architect and engineer Ithiel Town who developed this design and in 1820 took out a patent with the title "Town's Lattice Truss"[1]. The lattice trusses of the past had diagonal timber battens that criss-crossed between the top and bottom horizontal members and were screwed or nailed together. Based on small sections with identical dimensions, which were easy to produce, and nailed connections instead of time-consuming wood joints, these trusses could be easily assembled on site by untrained personnel. However, this form of construction faded into obscurity during the 20th century because it was too labour-intensive. Lack of experience regarding the fabrication and load-carrying capacity of

1 This economic, efficient form of construction was used for many covered bridges in the USA and Europe over the following decades. Even today, numerous lattice girder bridges are still with us. In the USA, Bull's Bridge in Connecticut, West Cornwall Bridge near Cornwall, Connecticut, and Eagleville Bridge in Washington County represent historic examples of this type of design.

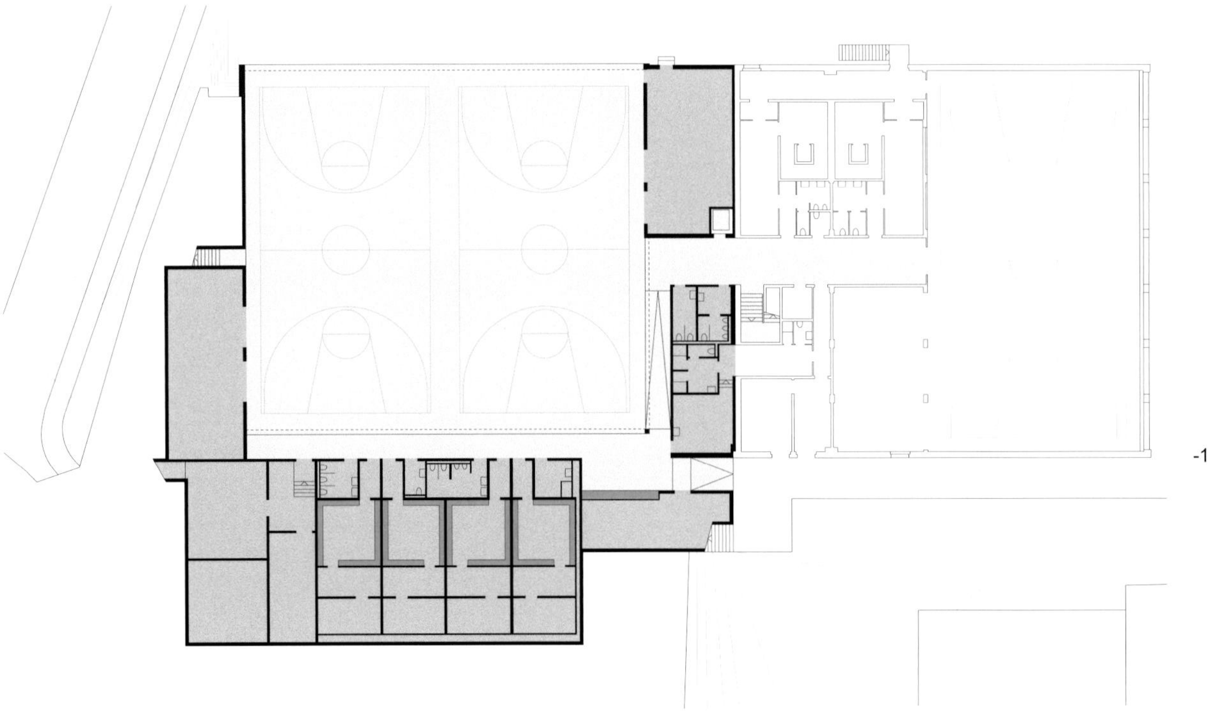

Level −1.

Spanning 32m, the lattice truss seems to float above the wide panoramic window.

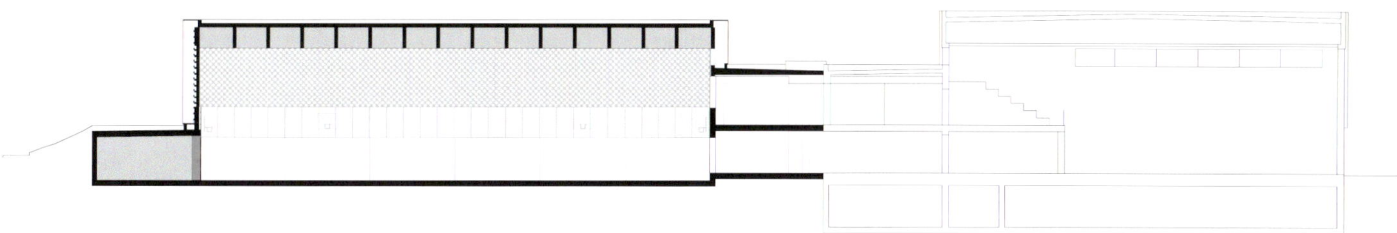

Longitudinal section.

this unusual form of timber construction led to the need to issue a second, Switzerland-wide tender before a timber contractor could be found who was ready to take on the challenge.

In contrast to their historical predecessors, the lattice trusses built for this sports hall consist of three layers of timber members. Different species of wood with strength classes between GL 24 and GL 36 as well as different cross-sections were chosen to suit the flow of forces. Between the 12cm wide × 1.20m deep top and bottom chords, which are glued laminated timber beams, are 120 × 120mm diagonals at 240mm centres in the same plane as the chords. Solid spruce sections measuring 12 × 40mm are attached over the full depth of each truss to function as diagonal ties. Likewise attached at 240mm centres, these ties are fixed to the

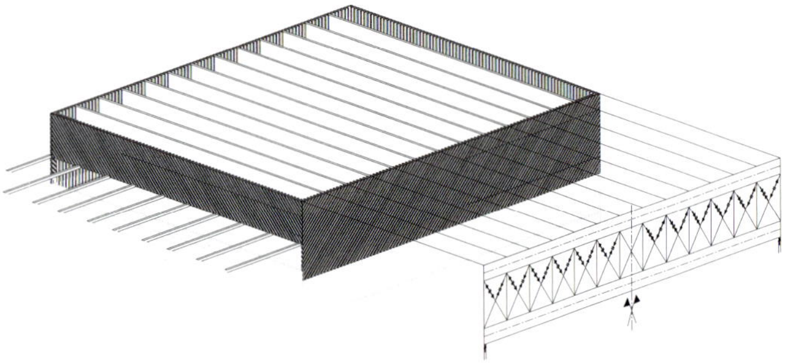

Isometric view of the structure.

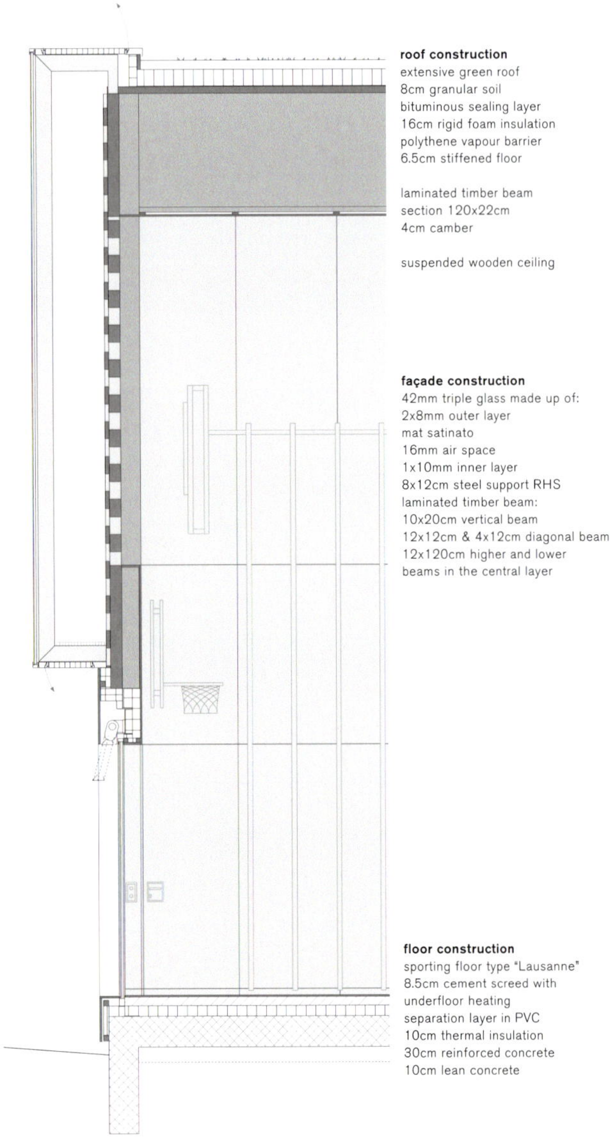

Vertical section.

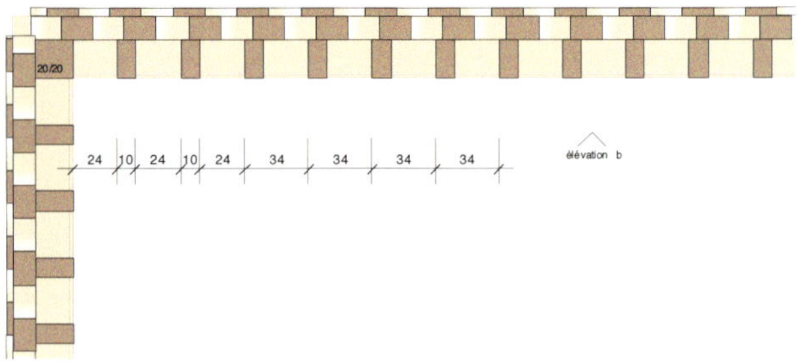

Horizontal section through loadbearing structure.

chords with tension-resistant nailed connections and to the diagonal struts by self-drilling screws. An additional, third, layer of vertical members is positioned on the inside of the trusses. Connected to the lattice by self-drilling screws, the 100 × 200mm glulam members at 340mm centres accommodate compressive forces.

Originally, the lattice trusses were to have been assembled on site like their historical predecessors. However, in order to guarantee the necessary precision of the almost 200m² trusses, they were factory-assembled and transported to the site 200km away as abnormal loads, where they were erected in just two weeks using two mobile cranes.

The 32m long lattice trusses carry a load of nearly 8t/m. Deformations must be kept to a minimum because of the adjacent glass façades. Roof and trusses are therefore rigidly connected together. The roof structure takes the form of 120 × 22cm glulam beams spanning across the playing area at a spacing of about 2m. The roof beams are rigidly connected to the top chords of the lattice trusses and braced by 65mm thick multiply boards.

Whereas in the interior the wooden surfaces of the beams and wall linings define the atmosphere, an envelope of acid-etched glass characterises the external appearance. Vertical, 6m high glass elements form a consistent, smooth surface to contrast with the resolute timber structure. The 42mm thick triple glazing is suspended about 1m in front of the timber lattice trusses.

On three sides, the 1.40m thick external walls of timber trusses plus glass outer leaf allow daylight to reach the playing area through almost the full height of

The diagonal members, loaded exclusively in compression, are connected to the horizontal top and bottom chords by way of a double oblique toe joint and are additionally secured by plates let into slits and dowels in double shear.

Plates let into slits strengthen the corner connections of each lattice truss; 1.5m wide and 1.5m high, they are fitted into the top and bottom chords of the truss.

the walls. The daylight, refracted by the etched glass, infiltrates through to the interior as diffuse illumination through the small openings in the lattice trusses and underpins the regular, non-directional appearance of the lattice trusses due to the multitude of criss-crossing members. From outside, the outline of the timber structure appears either blurred or distinct behind the façade, depending on the time of day and the available light. The superimposition and merging of the textures has turned the lattice trusses into more than just load-bearing elements; they have become both diaphanous external walls and decorative additions.

The accessible space between timber structure and glass façade contains automatically controlled ventilation flaps to ensure natural cross-ventilation in the sports hall.

Three roller-coasters

Colossos

Location	Heide Park 1, 29614 Soltau, Germany
Construction	9 months; completed: 2001
Design & structural engineering	Dipl.-Ing. Werner Stengel, Ing.-Büro Stengel, Munich
Fabrication drawings & production	Ing.-Holzbau Cordes, Rotenburg a. d. Wümme
Fabrication of timber rails	Merk Holzbau, Aichach, Germany

Balder

Location	Liseberg Park, Örgrytevägen 5, 40222 Gothenburg, Sweden
Construction	9 months; completed: 2003
Design & structural engineering	Dipl.-Ing. Werner Stengel, Ing.-Büro Stengel, Munich
Fabrication drawings & production	Ing.-Holzbau Cordes, Rotenburg a. d. Wümme
Fabrication of timber rails	Merk Holzbau, Aichach, Germany

Mammut

Location	Tripsdrill Theme Park, Treffentrill 1, 74389 Cleebronn, Germany
Completed	2008
Design & structural engineering	Dipl.-Ing. Werner Stengel, Ing.-Büro Stengel, Munich
Fabrication drawings & production	Ing.-Holzbau Cordes, Rotenburg a. d. Wümme
Fabrication of timber rails	Ing.-Holzbau Cordes, Rotenburg a. d. Wümme

Timber product used, structure	solid timber (split-heart pine)
Timber product used, rails	solid timber plies (pine, glued) or laminated veneer lumber

Simone Jeska

Mammut, Tripsdrill Theme Park: a roller-coaster that blends neatly into the landscape, despite its great height.

The growth of the theme park as a leisure and family attraction since its rediscovery in the 1970s has also brought about a renaissance in permanent timber roller-coasters[1].

The constantly changing ascending, descending, curved, inclined and straight sections of track correspond to a choreography of emotions and excitement defined by the circuit. Although each ride is designed individually to suit the local circumstances, and so results in a unique choreography, the supporting structures for roller-coasters essentially adhere to the same structural principles: a primary structure of trussed frames about 3m wide constructed from solid timber sections (6 × 14cm to 10 × 20cm) erected along the line of the track.

Trussed frames up to 60m high are needed to support the track at different levels. Longitudinal connections between the trussed frame sections, up to 11.60m long, take the form of butt joints and pairs of timber splice plates about 1m long connected with four bolts[2]. Diagonal and horizontal members are in the form of collars, i.e., pairs of members, connected via bolts and heavy-duty, low-deformation toothed-plate connectors[3] for tension- and compression-resistant connections to the vertical timber sections.

Longitudinal stability is ensured by joining the frames together with horizontal timber members (6 × 10cm, 4.5 × 10cm, 8 × 10cm), which are nailed in place. Cross-bracing (10 × 14cm members), which forms a diamond-shaped network right across the entire

1 It was in the late 19th century that the wooden slide, imported from Russia, underwent development in America and became the figure-of-eight circuit used for a whole series of roller-coasters. By the mid-20th century, demountable steel structures, which were erected and taken down again on fairgrounds in conjunction with travelling shows, were starting to replace the timber versions. Some 4000 timber roller-coasters have been built throughout the world since the start of the 20th century; currently, about five are built each year.

2 These connections must be able to handle the compressive and tensile forces caused by wind and dynamic loads; compression is transferred by direct bearing, tension by the splice plates.

3 Some 55 550 toothed-plate connectors were used for the Colossos in Soltau. Most of these were initially installed with an electric press tool and then subsequently tightened by hand or with a power screwdriver.

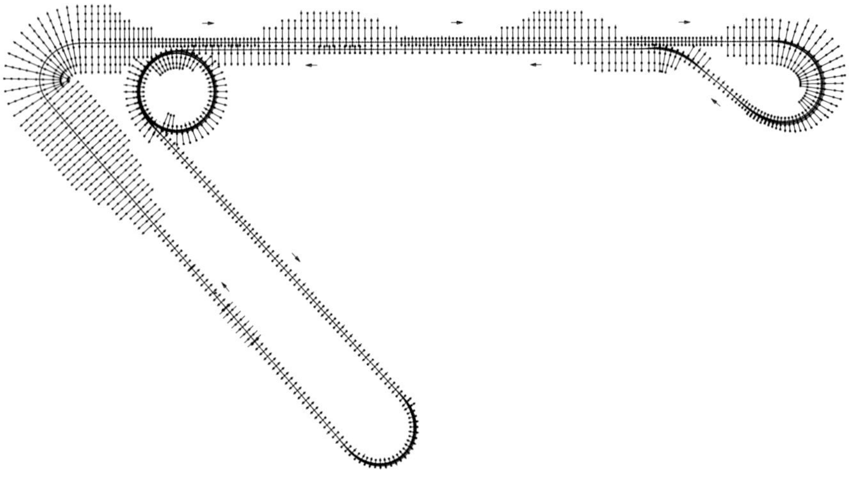

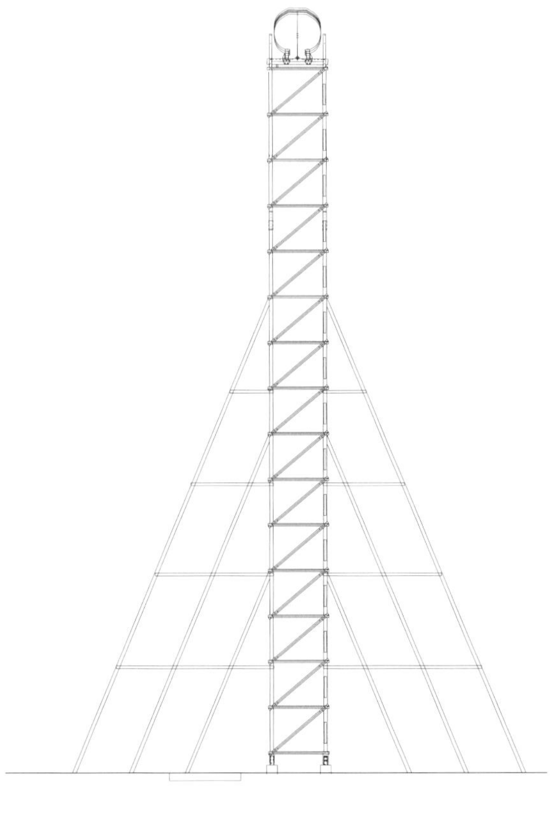

Plan of roller-coaster and principle of loadbearing structure for Colossos in Soltau, Germany.

The trussed frames are erected at spacings of 1.60–3.20m.

Horizontal timber members in the form of collars connect the inclined struts to the columns to brace the structure and reduce the buckling length.

The struts are joined together with nailed timber sections.

supporting structure, prevents buckling of the vertical frames. The bracing transfers the horizontal forces due to wind loads and the dynamic loads due to roller-coaster operation to the pad foundations. Where higher centrifugal forces or wind loads have to be resisted, the vertical frames are braced at different levels at intervals of about 7.50m on one or both sides by way of tension- and compression-resistant members, which in turn are connected together via members that are either horizontal or inclined to follow the flow of the forces. At the top of the vertical frames are the supports for the rails, the "ledgers" as they are called. As supports for the rails, the ledgers must not only be angled, but their top surfaces bevelled as well because the rails usually slope in two directions. The parallel rails, sometimes in double curvature, mounted on the ledgers are made from timber with a sheet steel running surface[4]; the double curvature results from the curving and at the same time inclined line of the track.

On straight sections of track the supporting structure is relatively uncomplicated. However, the multiple-curving, partly overlapping and multi-storey arrangement of a roller-coaster ride leads to a complex three-dimensional loadbearing structure that is assembled from a multitude of different, individual members. The ensuing extensive design, fabrication and erection workflows present challenges that go beyond just the logistics.

The introduction and coupling of digital design and production processes in the late 1990s simplified the fabrication and erection procedures associated with these huge timber structures and reduced the construction time enormously. In the meantime, the entire loadbearing structure can be depicted on the computer in three dimensions using just the plan layout and sketches of the structural system of the individual trussed frames. Individual drawings of the components with all their drilled holes and cutouts can then be produced auto-

The rails for Colossos in Soltau were prefabricated from LVL.

The rails, sometimes sloping in two directions and curved as well, are connected to the ledgers by steel framing anchors.

matically, converted to the technical data needed for fabrication, manufactured on five-axis CNC fabrication machines and labelled ready for erection[5]. A closer look at three timber roller-coasters, built within seven years of each other, helps to demonstrate the hurdles and opportunities of digital fabrication.

The availability of precise digital data enabled the rails for the timber roller-coaster in Soltau in northern Germany to be prefabricated in advance and erected at the same time as the loadbearing structure[6]. The cross-section of the rail bearers of laminated veneer lumber (LVL), in places in double curvature[7], is 20 × 40cm; this was cut out of larger sections in 6m lengths on five-axis CNC routers and fixed to the ledgers with hot-dip galvanised steel framing anchors (t = 12mm) with between two and five bolts per connection depending on the inclination of the rails.

During the design and fabrication of the timber roller-coaster in Soltau[8], all the drilled holes and cutouts

4 Sheet metal strengthens the rails in those areas of the track where the timber rails are loaded by the guide wheels of the cars. In addition, there is a nib on the inner face of each rail in some areas to prevent the cars from lifting off the rails.

5 Traditionally, all components were trimmed and drilled on site, which resulted in a colossal amount of work.

6 Engineer Werner Stengel specialises in the design of roller-coasters. In 1999 he applied for a patent for rails made from prefabricated, glued timber members mounted as complete rail segments. Erecting these at the same time as the loadbearing structure results in huge time-savings on site. Furthermore, the precise digital design and fabrication enables the course of the track to be adjusted to the "heart line". Using this heart line, the rails on curves are shifted in the opposite direction in order to avoid undesirable lateral acceleration of the cars as they traverse the track. Up until now, the rail bearers, made up of multi-ply, nailed timber sections, could not be erected until the entire loadbearing structure was finished. The bottommost ply was aligned (bent) manually on the ledgers and the other plies nailed to this. The sheet steel covering was added and adapted afterwards.

7 Compared with rails made from solid timber sections, the LVL bearers have a higher load-carrying capacity, better weather resistance (microcracks in the laminations ensure that the sections are fully impregnated), better dimensional stability and a longer life. The LVL bearers for Colossos were fabricated by Merk Holzbau in Aichach.

8 When it was built, the Colossos in Soltau was the largest timber roller-coaster in the world. The 1344m circuit rises to 51.40m and its steepest incline is 61°. The cars reach speeds of up to 110km/h.

Roller-coaster support structures have up to 90 000 timber members (many of which are one-offs in terms of their geometry and have skewed cutouts and inclined holes), which overlap and intersect each other to form what looks like a highly confusing structure.

plus the geometry of each curved member in the secondary structure had to be entered manually in the machine control for the CNC fabrication plant because of a lack of software interfaces[9]. Furthermore, the primary and secondary structures had to be designed separately and also broken down into segments in order to handle the huge quantity of data[10]. However, by the time work started on the roller-coaster in Gothenburg, the entire timber structure, including the secondary structure with its bracing, stairs, walkways alongside the track and safety barriers could be planned digitally as a whole[11]. Only the skewed, irregular halving joints and notches resulting from the overlaps and intersections of the secondary structure[12] had to be reworked manually at the machine control.

When it came to building the roller-coaster for Tripsdrill Theme Park near Cleebronn in southern Germany, new software developments enabled the stiffening effect of the rails to be included in the structural calculations for the first time. Furthermore, the unproblematic digital continuity from design through to fabrication plus the use of new connection technologies led to a new development in roller-coaster rails[13]. The rails – totally in keeping with tradition – were made from eight plies of 5cm thick pine sections connected with annular-ring shank nails and polyurethane glue[14], with all the work carried out on site. But in contrast to the traditional forms of construction, whose manual alignment caused irregularities and discontinuities in the line of the track, the position and alignment of the timber sections could be determined extremely precisely with the help of additional, digitally calculated and fabricated temporary members. Contour templates were spanned between the trussed frames to function as supports for additional, temporary ledgers[15]. Instead of the steel framing anchors used in the past, self-drilling SFS double-thread screws were used to connect the laminated rails to the ledgers.

The digitalisation of the whole procedure renders reworking of the timber members on site unnecessary – the components only need to be joined together. Clear labelling for assigning every individual component to its place within the entire structure plus an unambiguous assembly drawing are crucial here. As erection takes place successively from the bottom upwards and it is possible to climb the loadbearing structure at any point, the erection crew only needs attachment points for safety harnesses and cranes for lifting and positioning the components. The 90 000 timber components of the Colossos roller-coaster were assembled with millimetre accuracy over a period of seven months such that the deviations in level over a total height of about 52m were only max. 5mm. These small permissible tolerances are primarily due to the integration of the rails into the primary structure.

9 It was not possible to check the geometrical data on the computer controlling the fabrication plant owing to the incomplete graphical representation of the members and so some members still had to be trimmed on site.

10 The structural engineers produced 3000 drawings (system arrangements and details of individual members) for the primary structure to the roller-coaster in Soltau. In addition, the stiffening effect of the rails was not taken into account in the total system because the design program could not handle the huge number of individual members in the complete structure (90 000 members had to be designed). To take account of the fact that the rails distribute the loads (a considerable effect in track sections with high centrifugal forces), the track was modelled and designed in segments.

11 Despite "pruning" the files (no grid-lines, auxiliary elements or attributes), the 3D drawing of the entire structure, which comprises 34 000 individual members, exceeded a file size of 72MB and thus the capacity of the RAM available at that time. Software specialists had to be called in to combine the eight design files and integrate the faceted rail geometry into the 3D drawing.

12 The special feature of the "Balder" timber roller-coaster in Gothenburg is the spectacular track layout – in some places on three levels, one above the other, which is an outcome of the confined site. This layout resulted in additional overlaps and intersections in what was already a complex, bewildering loadbearing structure. The asymmetric arrangement of the inclined struts – again due to the lack of space – led to the need for elaborate foundations.

13 A patent with the title "Method for laying a timber rail for a fairground ride, especially for a roller-coaster, and timber rails for this purpose" was filed by the Ing.-Holzbau Cordes company in November 2007.

14 The glued joints were not taken into account when designing the rail bearers.

15 Separate contour templates are designed for the parallel rails because each rail has a different geometry, in particular where the track has a transverse gradient.

Digital 3D model of the timber roller-coaster in Gothenburg, Sweden.

The structural calculations for a roller-coaster structure are at least as complex as its fabrication and erection. Dynamic, short-term actions due to the motion of the cars, which cause additional horizontal loads particularly in the curves, must be taken into account in addition to the normal dead, wind and snow loads when designing the members and connections. Just the wind loads alone result in a theoretical horizontal displacement of about 10cm transverse to the track[16]. The different loads on the structure caused by dynamic actions and wind are reflected in the number of different connections (43 different node details for the roller-coaster in Soltau) and the design of the pad foundations, which must be designed accordingly for the specific loads of every column and strut.

Important aspects of timber structures permanently exposed to the weather are protection, corrosion of metal parts and deformations due to shrinkage and swelling. Although they entail considerable work on site, the small timber sections are advantageous from the point of view of swelling and shrinkage behaviour. Despite the enormous dimensions of the loadbearing structure, it merely becomes more elastic when affected by moisture. The use of pressure-impregnated, split-heart pine[17] in grades S 10 and S 13[18] minimises splitting and the infiltration of moisture. As the laminated rail bearers have to satisfy high demands with regard to dimensional stability, they are fully impregnated and also treated with a chemical preservative. To protect against corrosion, all steel connectors (bolts, anchors, nails, screws, etc.) are hot-dip galvanised and produced as custom articles; to allow for the thickness of the zinc coating, all bolts are produced undersize. Despite their permanent exposure to the weather, timber roller-coasters achieve a service life well in excess of 50 years; some timber roller-coasters in Europe have been in operation for nearly 100 years[19].

If we compare the loadbearing structures of the past with today's timber roller-coasters, then the influence of digital design and fabrication is obvious: the timber slides of the early days have become gigantic three-dimensional sculptures with spectacular track layouts.[20]

16 The actual displacement is, however, much smaller because the stiffening effect of the rails was not considered in the calculations; the rails in the curves function as a curved tie with a relatively large stiffening effect.

17 In split-heart timber, sawing the tree trunk down the middle avoids the stresses that start in the heartwood and thus minimises splitting. The timber is cut oversize, dried to impregnation moisture content, planed and pressure-impregnated only after fabrication.

18 Grades S 7 to S 13 designate the quality and strength of sawn timber. The use of solid timber in grades S 13 and S 10 means that only 25–30% of the trunk is utilised, because the pith has to be cut out of the trunk to produce timber free from heartwood (S 13). Cross-sections of grade S 13 and long components rendered necessary the felling of trees up to 140 years old.

19 Maintenance and monitoring of the structure principally involve checking the fit of the bolts, and regular inspections are carried out for this reason. A continuous ground slab over the entire area beneath the roller-coaster makes it easier to find any bolts that do drop out!

20 These days, different roller-coaster rides are compared by way of their "air-time", i.e., the passenger's feeling of weightlessness.

Travelling from a height of 60m at speeds of up to 100km/h is perceived as a free-fall experience. In the many curves and loops, over three storeys in places, the forces acting on the passengers reach up to four times their own body weight.

Toskana thermal baths

Bad Orb, Germany

Location	Horststraße 1, 63619 Bad Orb, Germany
Construction	2008; completed: May 2010
Architects	Ollertz Architekten, Fulda
Structural engineers	Trabert + Partner, Geisa
Fabrication drawings & production	HESS-Timber, Kleinheubach
Timber products used	glued laminated timber (spruce) combined with spruce battens

Simone Jeska

The new centre is a modern bathing landscape with therapy and healthcare areas to replace the old thermal baths dating from the 1960s.

Situated to the west of the spa gardens between the spa hotel and the salt graduation tower, the undulating roof over the new thermal baths in Bad Orb merges into the hills of the Spessart region. Corresponding to the free-form, terraced arrangement of the bathing landscape, the roof, a ribbed timber shell 65m long and 42m wide, covers the amorphous plan shape of the floor below, some 2200m² in area. In addition, the roof rises and falls in two directions with different amplitudes and wavelengths to produce a free-form geometry.

The starting point for the double-curvature roof geometry was a computer simulation of a suspended cable net fixed at eight points. Accordingly, the curves of the undulating roof follow the catenary curves that ensue with suspended cable nets. The geometric deviations in the form of bulges on the edge of the roof are due to simulating pretension in the cables. Further optimisation of the form was achieved with the help of the corresponding shell model to take account of the shear-resistant behaviour of the shell structure. Superimposing a diamond-shaped mesh matching the lines of the timber ribs on this leads to a shell structure of axially loaded compression members arranged on geodesic lines.

Superimposing the free-form roof geometry on the regular 1.80m grid results in various lengths between 0.19 and 3.80m for the individual glued laminated timber ribs (GL 24h), each of which spans over two bays. Rib cross-sections vary as well as rib lengths. Whereas the top surfaces of the glulam ribs follow the free-form

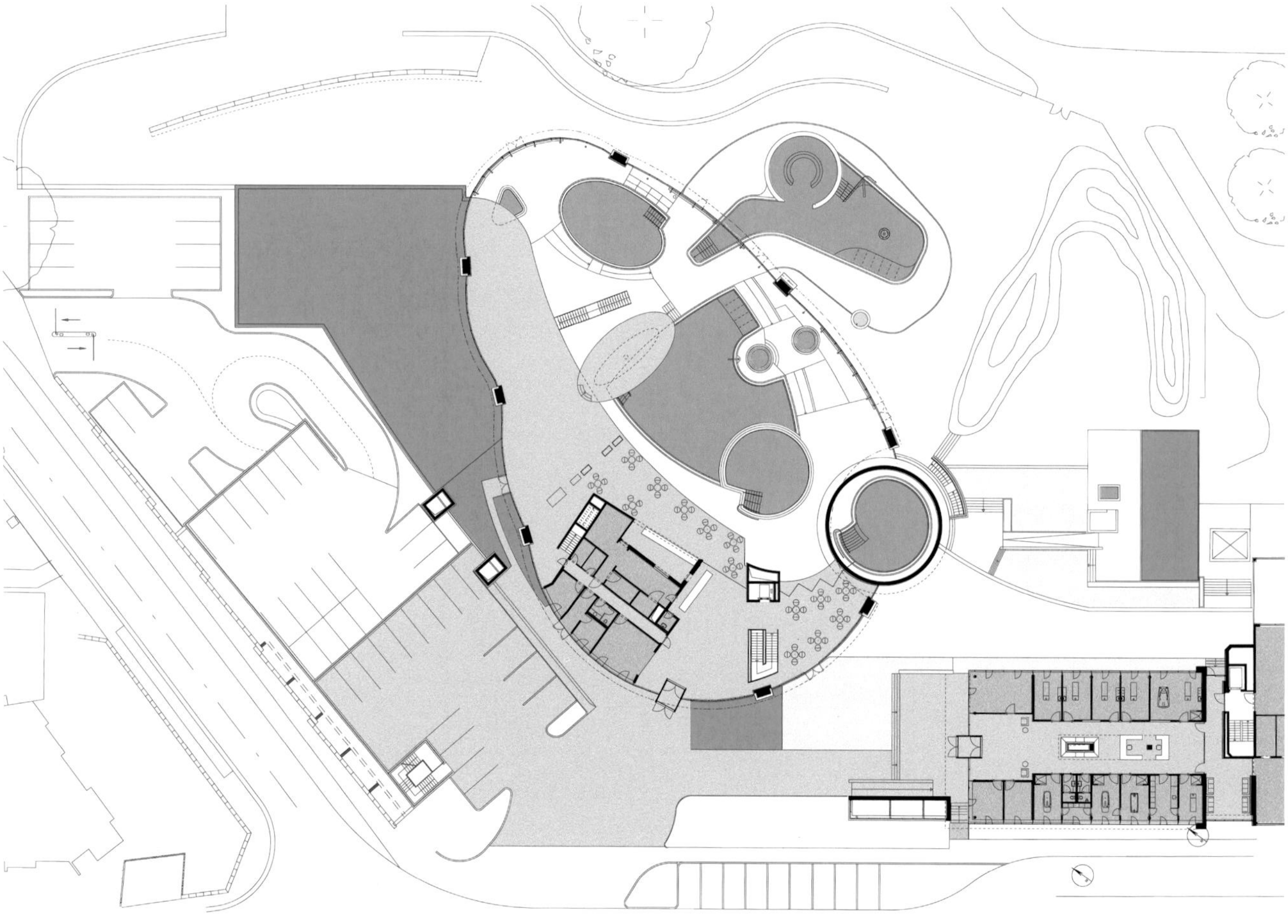

The plan form of the roof shell is made up of circular segments with different radii.

geometry of the roof surface, and have correspondingly twisted planes, the underside of each member traces one of 12 different predefined radii between 5 and 250m. Furthermore, each rib has a longitudinal 30 × 50mm rebate that enables the top edge of the rib to accommodate suspended acoustic elements on both sides. The different radii, lengths and surface geometries result in very complex, individual geometries for every one of the 682 timber ribs.

Along the undulating edge of the roof, the lattice shell is trimmed with a peripheral 240 × 800mm glulam beam (GL 28h, laid up in parallel blocks). This 170m long, double-curvature beam is made up of eight beam segments, each of which comprises 25 smaller pieces. At the low points of the edge beam, the roof is sup-

The curving roof leads to troughs reaching almost to the ground around the perimeter and crests rising to 10m high both internally and along the edge.

3D model of the roof structure.

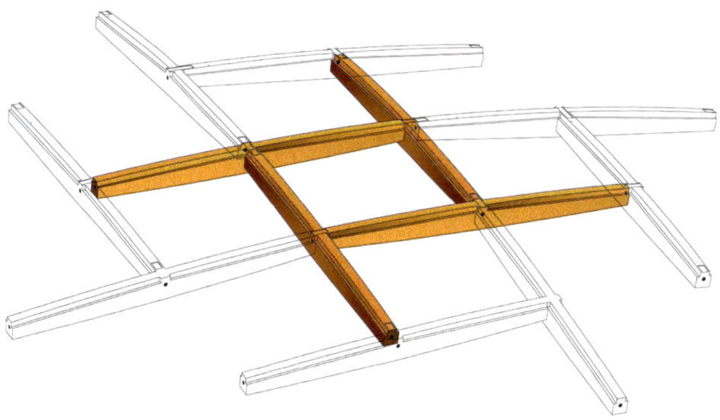

Using an idea borrowed from the Zollinger form of construction, the shell is made up exclusively of short members, each of which spans over two bays of the lattice.

ported on eight reinforced concrete plinths. On the eastern side, facing the spa gardens, the supports are only 2m high, on the western side they are positioned on the adjacent one-storey building.

The edge beam transfers the loads from the shell to the RC supports exclusively by direct bearing. To this end, these 2×0.8×2m blocks are provided with pockets in their top surfaces to match the double-curvature support geometry of the beam. The pocket geometries were determined digitally and the formwork fabricated on CNC woodworking machines.

Besides the structural calculations, the preparation of the data and the specifications required for fabricating the individual ribs represent a major challenge in the case of double-curvature free-form surfaces, and this can only be handled through digital continuity. In order to be able to assign the individual components unequivocally during erection, data transfers require not only the geometrical data, but also additional attributes such as component name or member number. The basis for the fabrication here was a 3D CAD model. This model enabled the preparation of individual com-

The clamping bed was specially developed with the help of digital techniques and produced with CNC routers in order to fabricate the double-curvature edge beam workpieces.

Slender internal steel columns at a regular spacing behind the façade prevent unwanted deformations and buckling of the edge at the junction with the glass façade.

ponent drawings with drilled holes, cutouts and production data.

To minimise the production costs, the glulam ribs for the lattice shell were initially allocated to 12 workpiece geometry groups with different radii. These curved workpieces, measuring 160 × 240mm in section, were processed on five sides on six-axis CNC machines and then numbered. Double-curvature workpieces (laid up in parallel blocks) were used for producing the edge beam segments, which needed single-curvature, 40mm thick loadbearing laminations made from 30 × 40mm spruce sections. Eight of these laminations with the corresponding curvature in the second plane were glued together under pressure in a clamping bed with resorcinol formaldehyde resin to form an oversize workpiece measuring 24 × 100cm. The precisely fabricated workpiece geometries meant that the double-curvature workpieces for the edge beam had to be machined on only four (narrow) sides and provided with

Numbering ensured unequivocal identification of the positions of the timber ribs within the total system.

The top and bottom surfaces of the curving glulam ribs have different radii of curvature.

the necessary cutouts, rebates, slots and drilled holes for the longitudinal joints as well as the connections between the ribs and the acoustic ceiling panels.

Approx. 40mm deep, individually formed cutouts enable the ribs to be connected to the edge beam. Depending on the angle at the junction, the ribs are either connected by screws (8 × 200mm) inserted at an angle of 45° with a punched metal plate fastener on the top, or are sawn to suit and connected to the edge beam with full-thread screws.

A special aspect of the design is the nodes between the ribs, which make use of traditional, hand-crafted wood joints. The curving timber members are merely

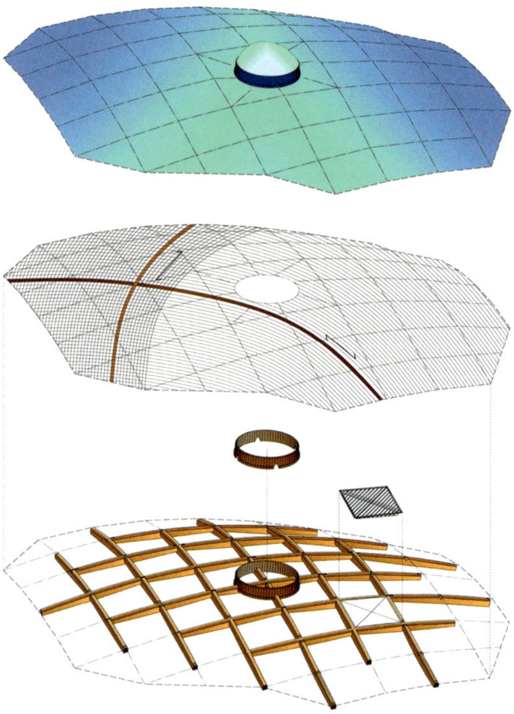

Two cross-wise plies of softwood boards, together with the lattice as the third layer, give the roof its form and make up the shell structure.

secured with 40mm diameter x approx. 14cm long beech dowels glued on one side and nailed steel plates.

This detail enabled the use of steel connectors, which are at risk of corrosion, to be reduced to a minimum – a crucial factor in a centre like this with its salt- and chlorine-laden atmosphere. Achieving such simple wood joints can, on the one hand, be attributed to efficient and accurate CNC fabrication, which allows holes to be drilled at various angles highly accurately, and, on the other, the adoption of a three-layer shell structure. The grid of ribs is the third layer, which together with the shear-resistant timber sheathing on the top of the shell, creates the shell geometry and stabilises the form by counteracting the structure's tendency to buckle under the effect of the loads. Theoretically, the shell response of the loadbearing structure is generated exclusively by the double layer of sheathing on the top because deformations due to shrinkage processes can allow gaps to form at the nodes between the glulam ribs, which prevent the compressive forces from being transferred. The timber sheathing of 30mm thick × 100mm wide softwood boards was nailed to the ribs and glued together with a polyurethane adhesive.

Starting at the edge beam, which was positioned accurately with the help of the concrete supports and

The free-form timber lattice shell was erected on falsework.

two further setting-out points, the glulam rib elements were joined one by one. The setting-out points defined the positions of the nodes, and orientation holes or marks on the ends of the members helped to align the individual ribs during erection. As a result of the accurately fabricated connections and member geometries, the complex roof geometry aligned itself between the setting-out points.

The extraordinary thing about this structure is not only the simplicity of the design of its free-form timber roof, but also the fact that with the full-height glazing, the remarkably thin roof shell seems to float above the landscape.

Whereas the beam segments near the supports are butt-jointed only, the segments between the supports are joined by plates let into the timber and dowels in multiple shear.

Inside the building, the undulating roof symbolises the waves of the water.

Index

Bibliography

GENERAL

Aicher, Simon; Reinhardt, H.-W.; Garrecht, Harald (eds.) (2013): Materials and Joints in Timber Structures: Recent Developments of Technology. Dordrecht, Heidelberg, New York: Springer.

Ernst, Hartmut (ed.) (2008): *Holzbau der Zukunft.* [Direction and coordination of the joint project: "Holzbau der Zukunft" (Wood construction of the future): TU Munich]. Stuttgart: Fraunhofer IRB Verlag.

Graubner, Wolfram (ed.) (1994): *Holzverbindungen. Gegenüberstellungen japanischer und europäischer Lösungen.* [Photos: Walter Grunder. Drawings: Louise Oldenbourg]. 5th ed. Stuttgart: DVA.

Haller, Peer (ed.) (2007): *Holzverbindungen und Holzverbundwerkstoffe. Fachbeiträge zum 7. Holzbauforum in Leipzig.* Berlin: Huss-Medien.

Haller, Peer (ed.) (2009): *Fertigung im Holzbau. Fachbeiträge zum 9. Holzbauforum in Leipzig.* Berlin: Huss-Medien.

Haller, Peer et al. (eds.) (2011): *Hochleistungsholztragwerke – HHT – Entwicklung von hochbelastbaren Verbundbauweisen im Holzbau mit faserverstärkten Kunststoffen, technischen Textilien und Formpressholz. BMBF-Vorhaben 0330722A-C; Abschlussbericht.* Dresden: TU Dresden, Institut für Stahl- und Holzbau (ISH) in cooperation with Institut für Textilmaschinen und Textile Hochleistungswerkstofftechnik (ITM).

Hascher, Rainer (ed.) (2009): *Grenzgänger aus Holz. 12. Fachtagung Holzbau Berlin und Brandenburg, 18. Dezember 2008.* Berlin.

Herzog, Thomas; Natterer, Julius; Schweitzer, Roland; Volz, Michael; Winter, Wolfgang (2004): *Timber Construction Manual.* 4th rev. ed., Basel, Munich: Birkhäuser, Edition Detail.

Holl, Christian; Siegele, Klaus (eds.) (2006): *Holz – große Tragwerke. Konstruktion, Architektur, Detail.* Baumaterialien series. Munich: DVA.

holz.bau forschungs GmbH Graz; TU Graz (2007): *Verbindungstechnik im Ingenieurholzbau. 6. Grazer Holzbau-Fachtagung.* Graz: Verlag der Technischen Universität Graz.

Kolb, Josef (2008): *Systems in Timber Engineering. Loadbearing Structures and Component Layers.* Basel: Birkhäuser.

Neuhaus, Helmuth (2009): *Ingenieurholzbau. Grundlagen, Bemessung, Nachweise, Beispiele.* 2nd ed. Wiesbaden: Vieweg+Teubner.

Niedermaier, Peter (2005): "Holz-Glas-Verbundkonstruktionen. Ein Beitrag zur Aussteifung von filigranen Holztragwerken". Doctoral diss. TU Munich.

Pfeifer, Günter; Liebers, Antje; Reiners, Holger (1998): *Der neue Holzbau. Aktuelle Architektur, alle Bausysteme, neue Technologien.* Munich: Callwey.

Schwaner, Kurt (ed.) (2009): *Zukunft Holz – Statusbericht zum aktuellen Stand der Verwendung von Holz und Holzprodukten im Bauwesen und Evaluierung künftiger Entwicklungspotenziale.* Biberach: Hochschule Biberach, Institut für Holzbau.

Steiger, Ludwig (2007): *Timber Construction.* Basel, Boston, Berlin: Birkhäuser.

Steurer, Anton (ed.) (2006): *Developments in Timber Engineering. The Swiss Contribution.* Basel, Boston, Berlin: Birkhäuser.

Stungo, Naomi (ed.) (1998): *Wood. New Directions in Design and Architecture.* Introd. Christoph Affentranger. San Francisco: Chronicle Books, 2001.

HISTORIC DESIGN TYPOLOGIES

Blaser, Werner (1982): *Schweizer Holzbrücken / Ponts de Bois en Suisse / Wooden Bridges in Switzerland.* Basel: Birkhäuser.

Blaser, Werner (1995): *Holz-Pionier-Architektur / Wood Pioneer Architecture. Meisterwerke seit 100 Jahren / Masterpieces of the last 100 years.* Weiningen, Zurich: Waser.

Blouet, G. Abel (ed.) (1834): *Supplemente zu Johann Rondelet's theoretischer und practischer Abhandlung über die Kunst zu bauen.* Leipzig, Darmstadt: Leske.

Brockstedt, Emil (1994): „Die Entwicklung des Ingenieurholzbaus am Beispiel der hölzernen Brücken im Zeitraum von 1800 bis 1940". Doctoral diss. TU Braunschweig.

Fraunhofer Informationszentrum Raum und Bau Stuttgart (ed.) (1991): *Kuppelbauten in Holzkonstruktion.* IRB-Literaturauslese, Vol. 2294. Editing by Folker Frank. 2nd ed. Stuttgart: Fraunhofer IRB Verlag.

Gattnar, Anton; Trysna, Franz (1961): *Hölzerne Dach- und Hallenbauten.* 7th ed. Berlin: Ernst & Sohn.

Gerold, Matthias (2001): *Holzbrücken am Weg. Einschließlich Geschichte des Holzbrückenbaus unter Berücksichtigung neuester Entwicklungen.* Karlsruhe: Bruderverlag.

Killer, Josef (1998): *Die Werke der Baumeister Grubenmann.* Ed. by Lignum, Schweizerische Arbeitsgemeinschaft für das Holz. 4th ed. Dietikon: Baufachverlag.

Krauth, Theodor; Meyer, Franz Sales (1993): *Das Zimmermannsbuch: Die Bau- und Kunstzimmerei mit besonderer Berücksichtigung der äußeren Form.* 11th reprint of 2nd improved ed., Leipzig: Seemann, 1895. Hanover: Schäfer.

Küttinger, Georg (1984): *Holzbau-Konstruktionen. Dachtragwerke, Hallen, Brücken.* Munich: Institut für Internationale Architektur-Dokumentation.

Lehfeldt, Paul (2001): *Die Holzbaukunst.* Holzminden: Reprint-Verlag Leipzig.

Pryce, Will (2005): *Architecture in Wood. A World History.* London: Thames & Hudson.

Seraphin, Mathias (2003): *Zur Entstehung des Ingenieurholzbaus. Eine Entwicklungsgeschichte.* Publication series of Lehrstuhl für Hochbaustatik und Tragwerksplanung, 2. Aachen: Shaker. Doctoral diss. TU Munich.

Steinmetz, Georg: *Grundlagen für das Bauen in Stadt und Land.* Editing by Deutschen Bund Heimatschutz. Munich: Callwey.

Wachsmann, Konrad; Grüning, Christa (1995): *Building the Wooden House. Technique and Design.* New edition. Basel: Birkhäuser.

Werner, Gerhard; Steck, Günter (1993): *Holzbau. Teil 2: Dach- und Hallentragwerke.* 4th ed. Dusseldorf: Werner.

Wesser, Rudolf (no date): *Der Holzbau.* Reprint of the 1903 Berlin edition. Holzminden: Reprint-Verlag Leipzig.

NEW TECHNOLOGIES AND METHODS

Antemann, Martin (2009): "Freiheit und Perfektion im Holzbau. Beispiel Dachkonstruktion Nine Bridges Yeoju Südkorea", In: *SAH 2009: Werkstoffkombinationen – ein Mehrwert für Holz*. Zurich: Lignum-Holzwirtschaft Schweiz. pp. 245–261.

Bejtka, Ireneusz (2005): *Verstärkung von Bauteilen aus Holz mit Vollgewindeschrauben* (Karlsruhe Report on Timber Engineering, Vol. 2). Doctoral diss. Universität Karlsruhe. Karlsruhe: Universitätsverlag Karlsruhe.

Blaß, Hans Joachim; Betjka, Ireneusz; Uibel, Thomas (2006): *Tragfähigkeit von Verbindungen mit selbstbohrenden Holzschrauben mit Vollgewinde* (Karlsruhe Report on Timber Engineering, Vol. 4). Karlsruhe: Universitätsverlag Karlsruhe.

Blaß, Hans Joachim; Romani, Markus (2000): *Trag- und Verformungsverhalten von Verbundträgern aus Brettschichtholz und faserverstärkten Kunststoffen*. Karlsruhe: Universitätsverlag Karlsruhe.

Blaß, Hans Joachim; Romani, Markus (2000): *Biegezugverstärkung von BS-Holz mit CFK- und AFK-Lamellen*. Karlsruhe: Universitätsverlag Karlsruhe.

Blaß, Hans Joachim; Romani, Markus; Schmid, Martin (2003): *Optimierung von Verbundträgern aus Brettschichtholz mit Verstärkungen aus Faserverbundkunststoffen*. Karlsruhe: Universitätsverlag Karlsruhe.

Burchardt, Bernd (1998): *Elastic Bonding: The basic principles of adhesive technology and a guide to its cost-effective use in industry*. Landsberg am Lech: verlag moderne industrie.

Endlich, Wilhelm; Steinmetz, E. (eds.) (1997): *Neue Entwicklungen in der Kleb- und Dichttechnik*. Essen: Vulkan-Verlag.

Forum-Holzbau (ed.) (2009): *15. Internationales Holzbau-Forum 2009*. Stuttgart: Fraunhofer IRB Verlag.

Forum-Holzbau (ed.) (2010): *16. Internationales Holzbau-Forum 2010*. Stuttgart: Fraunhofer IRB Verlag.

Frese, Matthias (2006): *Die Biegefestigkeit von Brettschichtholz aus Buche. Experimentelle und numerische Untersuchungen zum Laminierungseffekt* (Karlsruhe Report on Timber Engineering). Doctoral diss. Karlsruhe: Universitätsverlag Karlsruhe.

Göggel, Manfred (2000): *Bemessung im Holzbau*, Vol. 2: Verbindungen und Verbindungsmittel. 4th ed. Karlsruhe: Bruderverlag.

Habenicht, Gerd (2008): *Kleben – erfolgreich und fehlerfrei. Handwerk, Praktiker, Ausbildung, Industrie*. Wiesbaden: Vieweg+-Teubner.

Haller, Peer; Wehsener, Jörg (2003): *Entwicklung innovativer Verbindungen aus Pressholz und Glasfaserarmierung für den Ingenieurholzbau. AiF-Vorhaben 11164 B 1 von 1997 bis 1999*. Stuttgart: Fraunhofer IRB Verlag.

holz.bau forschungs GmbH Graz; TU Graz (2007): *Verbindungstechnik im Ingenieurholzbau. 6. Grazer Holzbau-Fachtagung*. Graz: Verlag der Technischen Universität Graz.

Kuhlmann, Ulrike; Brühl, Frank (eds.) (2010): *Holzbau Forschung + Praxis*. Doctoral research group. Universität Stuttgart. Institut für Konstruktion und Entwurf & Materialprüfungsanstalt Universität Stuttgart (Otto-Graf-Institut). Stuttgart: Universität Stuttgart.

Schäfers, Martin (2010): *Entwicklung von hybriden Bauteilen aus Holz und hochfesten bzw. ultrahochfesten Betonen. Experimentelle und theoretische Untersuchungen*. (Bauwerkserhaltung und Holzbau series, No. 4.) Kassel: Kassel University Press. [Doctoral diss. Universität Kassel, 2010].

Schwaner, Kurt (ed.) (2009): *Zukunft Holz – Statusbericht zum aktuellen Stand der Verwendung von Holz und Holzprodukten im Bauwesen und Evaluierung künftiger Entwicklungspotenziale*. Chapter 10: Verbindungstechniken. Biberach: Hochschule Biberach, Institut für Holzbau.

Uibel, Thomas; Blaß, Hans Joachim (2012): *Spaltverhalten von Holz beim Eindrehen von selbstbohrenden Holzschrauben*. Karlsruhe: Universitätsverlag Karlsruhe.

Wehsener, Jörg; Haller, Peer (2007): *Verstärkung von Holzverbindungen mit beanspruchungsgerechten textilen Strukturen. AiF-Vorhaben 14 ZBR*. Stuttgart: Fraunhofer IRB Verlag.

Articles

Brandmair, Anton; Steiner, Patrick; Niemz, Peter: "Klebstoffe bei tiefen Temperaturen verarbeiten", *Holz-Zentralblatt*, No. 5/2012, p. 152.

Fritzen, Klaus: "Kleben bei Holzverbindungen?", *Bauen mit Holz*, No. 3/2010, pp. 60–62.

Hübner, Ulrich: "Laubhölzer im Bauwesen sind noch Exoten", *Holz-Zentralblatt*, No. 37/2007, pp. 980–981.

Lehmann, Martin; Clénin, Reto; Richter, Klaus; Properzi, Milena: "Kohlefaserlamellen zur Verstärkung von Holzbalken", *Der Bauingenieur*, No. 10/2007, pp. 28–32.

Pörtner, Carsten; Seim, Werner: "In Holz eingeklebte stiftförmige faserverstärkte Kunststoffe", *Bautechnik*, Vol. 85, No. 4/2008, pp. 219–231.

Stahl, Jochen: "Transparenz, die trägt", *Bauen mit Holz*, No. 6/2010, pp. 28–30.

Steinmetz, Ralf; Lankes, Gottfried: "Holzturm mit aussteifenden Glasscheiben", *Bauen mit Holz*, No. 6/2004, pp. 16–25.

Tannert, Thomas; Hehl, Simon; Vallée, Till: "Probabilistische Bemessung von geklebten Anschlüssen im Holzbau", *Bautechnik*, Vol. 87, No. 10/2010, pp. 623–629.

Trautz, Martin; Koj, Christoph: "Mit Schrauben bewehren", *Bautechnik*, No. 3/2008, pp. 190–196.

Trinkert, Angela: "Geschwungen gebaut", *Bauen mit Holz*, No. 3/2008, pp. 14–19.

van de Kuilen, Jan-Willem; Torno, Stefan: "Esche für tragende Verwendungen. Festigkeitseigenschaften visuell sortierten Eschenholzes", *LWF aktuell*, No. 77/2012, pp. 18–19.

DEVELOPMENTS IN TIMBER CONSTRUCTION MATERIALS

Affentranger, Christoph (1997): *New Wood Architecture in Scandinavia*. Basel, Boston, Berlin: Birkhäuser.

Ambrozy, Heinz G.; Giertlová, Zuzana (2005): *Planungshandbuch Holzwerkstoffe. Technologie – Konstruktion – Anwendung*. Vienna, New York: Springer.

Bejtka, Ireneusz (2011): *Cross (CLT) and diagonal (DLT) laminated timber as innovative material for beam elements* (Karlsruhe Report on Timber Engineering, Vol. 17). Karlsruhe: KIT Scientific Publishing.

Benedetti, Cristina (2010): *Timber buildings. Low-energy constructions*. Bozen: Bozen-Bolzano University Press.

Benedetti, Cristina (2011): *Bauen mit Holz. Planungsdetails für Niedrigenergiegebäude*. 2nd ed. Bozen: Bozen-Bolzano University Press.

Deplazes, Andrea (ed.) (2008): Constructing Architecture: Materials, Processes, Structures. A Handbook. 3rd expanded ed. Basel: Birkhäuser.

Dietrich, Helmut; Glanzmann, Jutta; Knüsel, Paul; Sidler, Christine; Hegglin, Raphael; Humm, Othmar (eds.) (2012): *Holzbau – mehrgeschossig*. Zurich: Faktor Verlag.

Dworschak, Gunda (1999): *Der neue Systembau – Holz, Beton, Stahl. Skelett-, Tafel-, Zellenbauweise; Projektbeispiele, Konstruktion, Details, Kosten.* Dusseldorf: Werner.

Dworschak, Gunda; Wenke, Alfred (1998): *Holzvariationen. Mischkonstruktionen in Holz, Stein, Glas, Stahl.* Berlin: Verlag für Bauwesen.

Engel, Heino (2009): *Tragsysteme / Structure Systems.* 4. Aufl. Ostfildern: Hatje Cantz.

Fraunhofer Informationszentrum Raum und Bau Stuttgart (ed.) (1991): *Brettschichtkonstruktionen – Sporthallen, Mehrzweckhallen, Bäderbauten.* (IRB-Literaturauslese, Vol. 2444) Editing by Friedrich Breckner. 2nd ed. Stuttgart: Fraunhofer IRB Verlag.

Fraunhofer Informationszentrum Raum und Bau Stuttgart (ed.) (1994): *Leimverbindungen im Holzbau.* (IRB-Literaturauslese, Vol. 716) Editing by Theodor Löffler. 4th ed. Stuttgart: Fraunhofer IRB Verlag.

Fraunhofer-Informationszentrum Raum und Bau Stuttgart (ed.) (1995). *Brettschichtkonstruktionen – Industriehallen, Lagerhallen, Messehallen.* (IRB-Literaturdokumentation, Vol. 6842) Aktualisierte Fassung. Stuttgart: Fraunhofer IRB Verlag.

Gutdeutsch, Götz (ed.) (1996): *Building in Wood: Construction and Details.* Basel, Berlin, Boston: Birkhäuser.

Herrle, Peter (ed.) (2008): *Holzbau in der Stadt. 11. Fachtagung Holzbau Berlin und Brandenburg, 13. Dezember 2007.* Berlin.

Hugues, Theodor; Steiger, Ludwig; Weber, Johann (2004): *Holzbau. Details, Produkte, Beispiele.* 2. Aufl. München: Institut für Internationale Architektur-Dokumentation.

Ingenieurholzbau. Architektur, Tragwerk, Kosten (2004). Karlsruhe, Baden: Bruderverlag.

Ingenieurholzbau. Grundlagen, Bemessung, Nachweise, Beispiele (2011). 3rd ed. Wiesbaden: Vieweg+Teubner.

Kapfinger, Otto (ed.) (2009): *Hermann Kaufmann: Wood Works – Ökorationale Baukunst – Architecture durable.* Vienna, New York: Springer.

Kapfinger, Otto: Wieler, Ulrich (eds.) (2007): *Rieß Wood³ – Modulare Holzbausysteme.* Vienna, New York: Springer.

Kessel, Martin H. (ed.) (1998): *Bauen mit Holz und Holzwerkstoffen. Stand der Technik und Entwicklungstendenzen.* (Schriftenreihe Fraunhofer-Institut für Holzforschung), Wilhelm-Klauditz-Institut (WKI), WKI Report No. 33. Braunschweig: WKI.

Klinkenbusch, Claudia (2006): *Holzbauten der Moderne.* Niesky: Niesky Museum.

Kolb, Josef (2008): *Systems in Timber Engineering. Loadbearing Structures and Component Layers.* Basel: Birkhäuser.

Marboe, Isabella (ed.) (2010): *Holzbau. High performance in timber construction; Architects Collective.* Vienna, New York: Springer.

McLeod, Virginia (ed.) (2010): *Detail in Contemporary Timber Construction.* London: Lawrence King Publishing.

Moro, José Luis (ed.): *Baukonstruktion. Vom Prinzip zum Detail.* Berlin, Heidelberg: Springer.

Müller, Christian (ed.) (1998): “Entwicklung des Holzleimbaues unter besonderer Berücksichtigung der Erfindungen von Otto Hetzer. Ein Beitrag zur Geschichte der Bautechnik.” Doctoral diss. Bauhaus-Universität Weimar.

Müller, Christian (2000): *Laminated Timber Construction.* Basel, Berlin, Boston: Birkhäuser.

Schäfers, Martin (2010): *Entwicklung von hybriden Bauteilen aus Holz und hochfesten bzw. ultrahochfesten Betonen. Experimentelle und theoretische Untersuchungen.* Schriftenreihe Bauwerkserhaltung und Holzbau, 4. Kassel: Kassel University Press. Doctoral diss. University of Kassel.

Schleifer, Simone; Seidel, Florian (eds.) (2008): *Architecture materials – wood, bois, Holz.* Köln: Evergreen.

Spier, John (2006): *Building with engineered lumber.* Newtown, CT: Taunton Press.

Winter, Wolfgang (2005): *Holzbauweisen im verdichteten Wohnungsbau. Konstruktion, Bauphysik, Kosten.* Stuttgart: Fraunhofer IRB Verlag.

Zwerger, Klaus (2012): *Wood an Wood Joints. Building Traditions in Europe, Japan and China.* 2nd rev. and exp. ed., Basel, Berlin, Boston: Birkhäuser.

CNC PRODUCTION FOR TIMBER STRUCTURES

Antemann, Martin (2009): “Freiheit und Perfektion im Holzbau. Beispiel Dachkonstruktion Nine Bridges Yeoju Südkorea”. In: *SAH 2009: Werkstoffkombinationen – ein Mehrwert für Holz.* Zurich: Lignum-Holzwirtschaft Schweiz, pp. 245–261.

Beyer, Paul-Heinz (1991): *Technologie von CNC-Holzbearbeitungsmaschinen.* 2nd ed. Dusseldorf: Cornelsen/Schwann-Girardet.

Blumer, Hermann (2010): *Spektakuläres Konstruieren mit Holz.* Forum Holz I Bau I Frau. Meran 10. Stuttgart: Fraunhofer IRB Verlag.

Büren, Charles von: “Geflochten und Geformt”, *TEC21*, No. 7/2010, pp. 30–36.

Forum-Holzbau (ed.) (2009): *15. Internationales Holzbau-Forum 2009.* Stuttgart: Fraunhofer IRB Verlag.

Forum-Holzbau (ed.) (2011): *17. Internationales Holzbau-Forum 2011.* Stuttgart: Fraunhofer IRB Verlag.

Martynenko, Sergey; Heisel, Uwe (eds.) (2007): *Beitrag zur Steigerung der Effektivität und Prozesssicherheit der Hochleistungsbearbeitung von Holz- und Holzwerkstoffen mittels Bearbeitungszentren. Berichte aus dem Institut für Werkzeugmaschinen. Konstruktion und Fertigung.* Vol. 33. Stuttgart: Institut für Werkzeugmaschinen (IfW). [Doctoral diss. Universität Stuttgart.]

Nerdinger, Winfried et al. (eds.) (2010): *Wendepunkte im Bauen: Von der seriellen zur digitalen Architektur.* Munich: Edition Detail.

Scheurer, Fabian (2009): “Biegen statt Brechen – Komplexe Formen und digitale Prozesse im Holzbau”. In: Peer Haller (ed.), *Fertigung im Holzbau – Fachbeiträge zum 9. Holzbauforum Leipzig.* Berlin: Huss-Medien.

Scheurer, Fabian (2011): “Digitaler Workflow im Freiform-Holzbau”. In: *Forum Bois Construction Beaune 2011.* Biel: Forum Holzbau.

Schindler, Christoph (2009): “Auswirkungen der Informations-Werkzeug-Technik auf die Holzverarbeitung. Ein architektonisches Periodisierungsmodell anhand fertigungstechnischer Kriterien, dargestellt am Beispiel des Holzbaus”. Doctoral diss. ETH Zurich.

Schmid, Volker; Fischer, Markus (2010): *Metropol Parasol Sevilla – ein neues Wahrzeichen für den Ingenieurholzbau.* 16. Internationales Holzbau-Forum. Stuttgart: Fraunhofer IRB Verlag.

Articles

Antemann, Martin: “Freiheit und Perfektion im Holzbau”, *Bauen mit Holz*, No. 9/2009, pp. 8–12.

Blumer, Hermann: “Wie EDV-Anwendungen die Entwicklungen im Holzbau revolutionieren”, *Schweizer Holzbau*, No. 6/2007, pp. 26–31.

Bosse, Chris: "Digitales Origami – eine Installation in Sydney", *Detail*, No. 12/2007, pp. 1446.
Egle, Josef: "CAD-CAM-Schnittstellen im Holzbau", *Bauen mit Holz*, No. 3/1998, pp. 37–40.
Fischer, Martin: "Projekt 'Datentransfer im Holzbau' praxisreif", *Bauen mit Holz*, No. 4/1996, pp. 290–294.
Fritz, Oliver: "Handwerk im Computerzeitalter. CAD als Bindeglied zwischen Entwurf und Produktion", *Archithese*, No. 4/2006, pp. 26–31.
Fritzen, Klaus: "3-d und CNC-Fertigung. Mit Toleranzen und Prozeduren planen", *Bauen mit Holz*, No. 6/2003, pp. 16–17.
Garber, Richard: "Optimisation Stories. The Impact of Building Information Modelling on Contemporary Design Practice", *Architectural Design*, special issue: Closing the gap, No. 3–4/2009, pp. 6–13.
Gonchar, Joann: "Some Assembly Required", *Architectural Record*, No. 9/2008, pp. 138–146.
Haberle, Heiko: "Holzgeflecht aus 18 Kilometern Trägern", *Bauwelt*, No. 11/2010, pp. 32–33.
Hammerer, Reinhold: "Digitale Prozesse und Möglichkeiten der Individualisierung im Fertighausbau", *Detail*, No. 5/2010, pp. 488–489.
Hartmann, Heiner: "Multitalent Holzbau. Vielfalt der Konstruktionen", *Beratende Ingenieure*, No. 5/2005, pp. 10–14.
Hovestadt, Ludger: "Die technische Rekonstruktion der Architektur im Informationszeitalter", *Detail*, No. 12/2007, pp. 1434–1438.
Kleilein, Doris: "Burst 003", *Bauwelt*, No. 15/2008, pp. 30–33.
Klein, Stephan: "Centre Pompidou. Ein Dach wie eine Achterbahn", *Mikado*, No. 2/2010, pp. 34–36.
Krogmann, Hubert: "Dachkonstruktion nach historischer Bauart". *Detail*, No. 1/2000, pp. 102.
Leopold, Bernd; Egert, Bernd: "Eissporthalle in St. Pölten – Tragwerk, Bauphysik, Haustechnik. St. Pölten Ice Rink – Frame, Physics, Installations", *Detail*, No. 7–8/2008, pp. 810–816.
Maier-Solgk, Frank; Redecke, Sebastian: "Abenteuer Centre Pompidou. Die Dependance in Metz", *Bauwelt*, No. 22/2010, pp. 6–9.
Menges, Achim: "Architektonische Form- und Materialwerdung am Übergang von Computer Aided zu Computational Design", *Detail*, No. 5/2010, pp. 420–425.
Meyhöfer, Dirk: "Metropol Parasol in Sevilla. Raum für urbanes Leben des 21. Jhd.", *DBZ*, No. 3/2011, pp. 22–27.
O'Grady, Sandra Kaji: "BURST * 003", *AA*, No. 09–10/2006, pp. 80–87.
Pawlitschko, Roland: "Metropol Parasol – Stadtlandschaft in Sevilla", *Detail*, No. 12/2007, pp. 1444–1445.
Schindler, Christoph: "Die Mittel der Zeit – Herstellungsinnovation im Holzbau", Arch+, No. 188/2008, pp. 92–95.
Steuerwald, Thomas: "Zollinger-Bauweise. Messedach übernimmt tragende Rolle", *Mikado*, No. 9/2002, pp. 16–20.

EXPERIMENTAL AND TEMPORARY STRUCTURES

Battisti, Valentin (2009): *Faltwerke aus Brettsperrholz.* Doctoral diss. TU Graz. Graz: Institut für Holzbau und Holztechnologie.
Leitner, Katharina (2004): *Tragkonstruktionen aus plattenförmigen Holzwerkstoffen mit der textilen Fuge.* Aachen: Verlagsgruppe Mainz.
Weinand, Yves (2011): "New Structural Potential of Wood: The Ibois Research Laboratory at EPF Lausanne". In: Flury, Aita (ed.), *Cooperation. The Engineer and the Architect.* Basel, Boston, Berlin: Birkhäuser, pp. 91–101.

Articles

Adam, Hubertus: "Gesegnetes Provisorium. Kapelle in Pompaples", *db*, No. 9/2010, pp. 30–34.
Buri, Hani; Weinand, Yves: "Gefaltet", *TEC21*, No. 8/2009, pp. 18–21.
Buri, Hani; Weinand, Yves: "Origami aus Brettsperrholz", *Detail*, No. 10/2010, pp. 1066–1068.
Carl, Seraina: "Temporäre Kapelle von Saint-Loup, Schweiz – von Faltwerken inspiriert", *Architektur.aktuell*, No. 11/2009, pp. 80–91.
Führer, Wilfried; Leitner, Katharina: "Innovative Faltwerke aus Holzwerkstoffen", *Beratende Ingenieure*, No. 5/2000, pp. 36–39.
Geipel, Kaye: "Das elementare Holzhaus", *Bauwelt*, No. 39–40/2009, pp. 0–43.
Gunßer, Christoph: "Material ist teuer, Geometrie ist billig", *db*, No. 9/2010, pp. 70–71.
Hovestadt, Ludger; Mueller, Dennis: "ESG-Pavillon – Digitale Technologien beim Entwerfen und Produzieren", *Detail*, No. 12/2004, pp. 1484–1487.
Kaltenbach, Frank: "Teaching by Doing – Forschungspavillon in Stuttgart", *Detail*, No. 10/2010, pp. 994–995.
Kaltenbach, Frank: "Urhütte der Zukunft? Bionischer Forschungspavillon aus Holz", *Detail*, No. 1–2/2012, pp. 6–8.
Kohlhamer, Thomas: "Holzbausteine. Ein individuell veränderbares Wandsystem", *Baumeister*, No. 9/2009, pp. 72–73.
Leitner, Katharina: "Faltwerke mit 'Textiler Fuge'. Konstruktion aus gefalteten Holzwerkstoffen", *Bauen mit Holz*, No. 5/2003, pp. 30–33.
Lienhard, Julian; Fleischmann, Moritz; D'Souza, Diana; Schleicher, Simon: "Biegung erwünscht", *Bauen mit Holz*, No. 2/2011, pp. 30–34.
Lucan, Jacques: "Fundstücke", *Werk, Bauen + Wohnen*, No. 10/2005, pp. 4–8.
Müller, Joachim: "Bionisch inspirierte Materialsysteme. Dem Genie der Natur auf der Spur", *Umrisse – Zeitschrift für Baukultur*, No. 4/2009, pp. 30–34.
Nabaei, Sina; Weinand, Yves: "A modular timber structure. Parametric design as a feature for structural improvement", *Holzforschung Schweiz*, No. 1/2012, pp. 12–15.
Schärer, Casper: "Origami in Holz", *Werk, Bauen + Wohnen*, No. 5/2009, pp. 12–17.
Schindler, Christoph: "Das neue Bild vom Holz – Digitale Holzbearbeitung zur Umsetzung gekrümmter Formen", *Detail*, No. 11/2008, pp. 1310–1316.
Thönnissen, Udo; Werenfels, Nik; Spiro, Annette: "Hebelstabwerke in Forschung und Lehre", *Holzforschung Schweiz*, No. 2/2011, pp. 6–9.
Trautz, Martin: "Das Prinzip des Faltens. Folded-Plate Principles", *Detail*, No. 12/2009, pp. 1368–1376.
Weinand, Yves: "Origami, Brettrippen und Reibschweissen", *Der Bauingenieur*, No. 5/2006, pp. 60–63.

Illustration credits

Ambrosetti, Tonatiuh, ETH-Studio Monte Rosa 71
Architectural Association Photo & Film Library 79 l
Arup Ingenieure 49, 75 l
Atelier Peter Zumthor & Partner 77 al
Augustin, Manfred, hbf Graz 15 r
aus: „Versuche mit eingeleimten Gewindestangen", Karl Möhler, Klaus Hemmer, Universität Karlsruhe, 1981, Fraunhofer IRB Verlag 17
Baan, Iwan 77 br
Bathon, Leander, Labor für Holzbau, HS RheinMain 20 b
Bejtka, Ireneusz, Karlsruhe 15 l, 27 r
Berner Fachhochschule Biel, AHB 26
Bletz, Oliver, Labor für Holzbau, HS RheinMain 20 a
Bockhop, Svenja 95
Borgmann, Roland, Münster 87
Bosch AG 39
Burgbacher Holztechnologie GmbH 18
Büro für Ingenieur-Architektur, R. J. Dietrich 130, 131 r, 132, 134
Cordes Holzbau 153 a, 154, 155
designtoproduction 67, 68 r, 69 r, 74 a, 112, 113 br
Dietrich, Richard J. 129, 131 l, 133
Döbele, Tobias 66, 69 l
Eckhardt, Christian, Evonik Industries AG 35 b
Edmiston, Jeremy; Gauthier, Douglas 62, 72, 73
Erlebnispark Tripsdrill GmbH & Co. KG 65 a, 151, 153 b, 158
Esters, Eckehart 88 b, 88 ar
Flechtner, Thomas 77 bl
Franke, Peter, Punctum Fotografie 65 b, 159, 161 a, 162 b, 166 b
Golinski, Markus 25 a
Gonoso, Gonzalo 106, 108 bl, 108 br
Graeme Mann & Patricia Capua Mann 146, 147 b, 148, 149
Gramazio & Kohler, ETH Zürich 84, 86
Hacke, Mila, Berlin 78 r
Haller, Peer, TU Dresden, Institut für Stahl- und Holzbau 28, 29
Hans Hundegger Maschinenbau GmbH 60 r, 63
Häring & Co. AG 19, 32
Hascher, Jehle und Assoziierte GmbH 96, 97 b
Hasenauer, Christian, Design bei Superlab / www.superlab.at 35 a
Hasslacher Norica Timber 120 r
Herzog, Alain, EPFL 90 l
HESS TIMBER 25 b, 162 a, 163 a, 165, 166 a
Hochhauser, Werner 34
Hudert, Markus, IBOIS, EPFL 83
Hurnaus, Hertha 61
Hurtado, Martín 36, 37 r
IBOIS, EPFL 85 a
ICD/ ITKE Universität Stuttgart 76, 80, 82, 91, 92
Ingenieurbüro Stengel 152 l
Ingenieurbüro Trabert + Partner 22, 161 b, 164 a
Inros Lackner AG 13 al, bl
Jantscher, Thomas, Colombier 145, 147 a, 150
Jeska, Simone 152 r
Jong-Oh, Kim 68 l, 109, 114, 116 r
Kaulquappe 140 al, 142 l
Keller, Milo 89 r
Keller, Roman, Zürich 14, 31, 85 b
Knechtl, Christian 121
Könsgen, Matthias 97, 99 ar, 99 br, 100
Kronoply GmbH 50 r
Kyeong Sik Yoon/KACI International 110, 111 b
Labor für Holzbau, HS RheinMain 23
Leiska, Heiner 11 b
Leitner, Kathrin 88 al
Lignotrend® Produktions GmbH 57 b
Localarchitecture 89 l
Loebermann, Matthias, Hochschule Biberach, IKB 78 l
LTC Lehmann timber construction 59, 111 a, 113 a, 113 bl, 115, 116 al, 116 bl
Markus Schietsch Architekten 138, 139 a, 140 bl, 141 r, 143, 144 a
Martinez, Ignacio 51
Meyer, Jens, BUGA GmbH 135
Mozó, Alberto 102, 103 b, 104, 106 a, 108 a
Nabaei, Sina, IBOIS, EPFL 90 r
Neue Holzbau AG, Lungern 21, 30
Ollertz Architekten 160, 163 b, 164 b
Pinzón, Nicolás 45
Reichert, Steffen, ICD, Uni Stuttgart 93
Rutter, David, Berlin 79 r
Saleh Pascha, Khaled 38, 42, 43, 46, 47, 48, 50 l, 53, 54, 57, 101, 105 bl, 105 br, 107, 119, 120 l
Schaffitzel Holzindustrie GmbH 136
Schanda, Irene 117, 118, 122
Schmid, Volker, Arup Ingenieure 75 r
Schrentewein, Thomas 123, 124, 125, 126, 127, 128
Schürmann, Helmut, TU Darmstadt 24
Schweizer Rheinsalinen AG 40
Sou Fujimoto Architects 77 ar
Sprenger, L. 33
Steiner, Patrick, Fa. Nolax AG 16
Trautz, Martin, RWTH Aachen 27 l
Tyler, Adrian 44 l, r
Universität Stuttgart 60 l
Uusheimo, Tuomas 58, 74 b
Walt + Galmarini AG 137, 139 b, 140 r, 141 l, 142 r, 144 b
Weinand, Yves, IBOIS, EPFL 55 r
WIEHAG GmbH 98, 99 l
Wulff, Heiko 157

About the authors

Rainer Hascher, born in 1950, studied architecture at the University of Stuttgart and founded his own architectural practice in 1979. Following several professorial appointments at the universities in Aachen, Stuttgart and Berlin, he taught structural design and climate-friendly building at Technische Universität Berlin from 1989 to 2013. The collaboration with Sebastian Jehle began in 1992. Besides successes in many competitions, the Hascher Jehle architectural practice has received numerous awards for completed projects, including the German Architecture Prize, the German Urban Planning Prize and the German Steel, Concrete and Timber Design Prizes. He was appointed to the Berlin-Brandenburg Academy of Sciences and Humanities in 2001. The Hascher Jehle practice took part in the 10th International Architecture Exhibition in Venice in 2006. He was a founding member of the German Sustainable Building Council (DGNB) in 2007.

Simone Jeska, born in 1965, studied architecture at Technische Hochschule Nuremberg and the theory and history of architecture at ETH Zurich. She lives in Berlin and works as a scientific assistant at the Institute of Architecture, Technische Universität Berlin. As a freelance author, she has already published several books on architecture, including *Office Buildings: A Design Manual* and *Transparent Plastics: Technology and Design* plus an architectural guide for children in German.

Khaled Saleh Pascha, born in 1965, studied architecture at Technische Universität Berlin, where he also earned his doctorate. Six months of the year he is a professor at the Catholic University in Santiago de Chile, the other six months a post-doctoral university assistant at Vienna University of Technology. His research and teaching activities embrace timber construction, bioclimatic architecture and architectural theory. He is the author of diverse articles and contributions to books on these subjects which have been published in the UK, Austria, Germany and Chile.

Acknowledgements

For the many informative discussions and for their help and assistance, we would like to thank Hermann Blumer, Christian Burgbacher, Charles von Büren, Ulf Cordes, Mathias Hofmann, Hans Hundegger, Christian Knechtl, Wolfram Kübler, Alberto Mozó, Fabian Scheurer, Joachim Schindelhauer, Thomas Schrentewein and Josef Trabert.
We thank Julia Pauli for her exceptional dedication and tireless efforts in researching images.
And, last but not least, we extend special thanks to our editor Henriette Mueller-Stahl.